SpringerBriefs in Applied Sciences and Technology

PoliMI SpringerBriefs

Series Editors

Barbara Pernici, DEIB, Politecnico di Milano, Milano, Italy

Stefano Della Torre, DABC, Politecnico di Milano, Milano, Italy

Bianca M. Colosimo, DMEC, Politecnico di Milano, Milano, Italy

Tiziano Faravelli, DCHEM, Politecnico di Milano, Milano, Italy

Roberto Paolucci, DICA, Politecnico di Milano, Milano, Italy

Silvia Piardi, Design, Politecnico di Milano, Milano, Italy

Gabriele Pasqui ⓘ, DASTU, Politecnico di Milano, Milano, Italy

Springer, in cooperation with Politecnico di Milano, publishes the PoliMI Springer-Briefs, concise summaries of cutting-edge research and practical applications across a wide spectrum of fields. Featuring compact volumes of 50 to 125 (150 as a maximum) pages, the series covers a range of contents from professional to academic in the following research areas carried out at Politecnico:

- Aerospace Engineering
- Bioengineering
- Electrical Engineering
- Energy and Nuclear Science and Technology
- Environmental and Infrastructure Engineering
- Industrial Chemistry and Chemical Engineering
- Information Technology
- Management, Economics and Industrial Engineering
- Materials Engineering
- Mathematical Models and Methods in Engineering
- Mechanical Engineering
- Structural Seismic and Geotechnical Engineering
- Built Environment and Construction Engineering
- Physics
- Design and Technologies
- Urban Planning, Design, and Policy

Sabrina Bresciani

Editor

Social Innovation Projects for Climate Neutral Cities

Making Municipalities Sustainable
with People-Based Solutions

POLITECNICO
MILANO 1863

Editor
Sabrina Bresciani
Department of Design
Politecnico di Milano
Milan, Italy

ISSN 2191-530X   ISSN 2191-5318   (electronic)
SpringerBriefs in Applied Sciences and Technology
ISSN 2282-2577   ISSN 2282-2585   (electronic)
PoliMI SpringerBriefs
ISBN 978-3-031-87725-4   ISBN 978-3-031-87726-1   (eBook)
https://doi.org/10.1007/978-3-031-87726-1

This work was supported by Politecnico di Milano.

This Springer imprint is published by the registered company Springer Nature Switzerland AG
The registered company address is: Gewerbestrasse 11, 6330 Cham, Switzerland

If disposing of this product, please recycle the paper.

Preface

Achieving climate neutrality is an ambitious goal for cities worldwide that has positioned social innovation as a critical lever for systemic urban change toward sustainability. This open access book provides cases, methods, pathways, and strategies of multi-stakeholder engagement and people-based actions that municipalities can deploy for reducing greenhouse gas (GHG) emissions and improving sustainable urban living. The eight chapters provide an overview of social innovation's ability to contribute to achieving net-zero targets in urban contexts, drawing on empirical findings and case studies developed within the NetZeroCities project. This large EU-funded project seeks to support European cities in their ambition to become climate neutral, in particular providing continuous and tailored support to over 100 cities to reach net zero by 2030.

In the context of urban climate action, social innovation goes beyond conventional techno-centred approaches by supporting the establishment and scaling of collaborative solutions developed by diverse stakeholders within cities: residents, civic organizations, municipal authorities, and private entities. Beyond engagement, social innovations in urban governance comprise actions to increase society's capacity to act, for example by build platforms for knowledge sharing, co-designing, and co-financing solutions. This approach acknowledges cities not simply as administrative units but as the hubs that facilitate systemic co-produced interventions that can produce radical impacts not only in reducing emissions but also increasing democratization, inclusion, and wellbeing.

Structured in eight chapters written by a diverse range of scholars and experts involved in supporting European cities to reach climate neutrality by 2030, the book presents a series of methodologies and case studies that illustrate how social innovation can be purposefully embedded in city climate action plans, based on the NetZeroCities project. The first chapter frames the role of social innovation in achieving climate neutrality, proposing it as a key lever for reaching the scale of innovation required to achieve such ambitious goal as climate neutrality, by activating the urban ecosystem of private companies, governmental, and civic organizations as well as citizens, for co-creating solutions that are social and environmental in the means and in the ends. The second chapter focuses on the development of robust urban ecosystems to support social innovation, detailing methods to organize and sustain collaborative urban networks. These frameworks provide an institutional basis for cities to incorporate social innovation across various domains of urban policy, ensuring that GHG reduction efforts are supported by sustained social engagement.

Building on these foundations, the following chapter explores how cities can integrate social innovation within their climate action portfolios: the third chapter outlines the mechanisms through which cities can make social innovation relevant and actionable in the context of climate neutrality. By adopting strategic frameworks that emphasize relevance and scalability, cities can move from experimental projects to institutionalized practices, thereby enhancing their capacity for long-term climate action. This chapter's insights are especially relevant for city leaders and policymakers who seek to align social innovation initiatives with broader urban climate goals.

The fourth and fifth chapters provide a detailed analysis of social innovation pathways that are both actionable and scalable. In particular, the fourth chapter examines the design of scalable social innovation models, visually presenting ten categories of interventions that cities can adopt based on specific local needs and resources, from the preparatory phase, to action and scaling. The fifth chapter a complementary presentation of ten case studies covers three key urban challenges: citizen engagement, energy transition, and behavioral change. These cases illustrate how social innovation practices are not add-ons to climate policies but essential drivers of behavior and local engagement. By centering the human dimension in climate action, these interventions have achieved measurable impacts in GHG reduction, showcasing the viability of social innovation as a climate solution in diverse urban contexts.

The book also explores the operational and governance structures necessary to support social innovation at the city level. The sixth chapter presents the Wiener Klimateam project as a case study, offering insights into how cities can cultivate civic ecosystems conducive to social innovation. Through mechanisms that strengthen civic *soil*, such as participatory budgeting and community-led initiatives, this chapter demonstrates—through the case of the city of Vienna—how cities can create enabling environments for grassroots innovations. This perspective emphasizes that social innovation does not exist in isolation but relies on institutional and policy frameworks that recognize and facilitate civic agency.

Methodologically, this book provides an array of tools for practitioners and researchers to understand and replicate social innovation processes in their respective cities. The seventh chapter introduces methods that can be utilized for supporting the diverse phases of social innovations development, outlining through an interactive visual a series of frameworks that guide cities through the stages of ideation, prototyping, and implementation of social innovation initiatives. The variety of methods provided takes stock of the broad literature on social innovation methods to allow cities to select and customize approaches based on contextual challenges and opportunities. The chapter underscores the importance of iterative design in social innovation, where solutions are continuously refined in response to feedback and evolving urban dynamics.

Finally, the eighth chapter situates social innovation within the framework of the New European Bauhaus, proposing a portfolio-based approach to foster climate-neutral urban futures. This perspective connects social innovation with urban aesthetics and sensorial aspects, suggesting that urban environments can be purposefully crafted to support both environmental goals and social well-being.

Beyond the case studies and methodologies, this book makes a significant academic contribution by providing a comprehensive and multi-faceted analysis of social innovation's role in urban climate action, offering a wide-ranging examination of strategies, pathways, methods, literature, and case studies that are relevant to scholars, policymakers, and practitioners. By linking social innovation with systemic approaches to climate neutrality, it provides a foundation for future research on the intersection of urban governance, civic and multi-stakeholders engagement for environmental sustainability, supporting and enhancing technical climate actions. Moreover, it shows that social innovation solutions are successfully deployed in diverse geographical contexts, illustrating solutions practiced by cities across Europe and the World.

Through this book, we aim to contribute to an academic and practical discourse that positions social innovation not as a peripheral approach to urban governance and climate neutrality but as an integral and necessary lever for systemic innovation in climate action.

Milan, Italy   Sabrina Bresciani

Acknowledgments

The work presented in this book was conducted as part of the project "NetZeroCities" which received funding from the European Union's Horizon Europe research and innovation programme under grant agreement No. 101036519 (H2020) and No. 101121530—SGA-NZC (Horizon Europe) Website: https://netzerocities.eu/

The research was funded by the European Union. However, the views and opinions expressed are solely those of the authors and do not necessarily represent those of the European Union or CINEA. Neither the European Union nor the granting authority can be held responsible for them.

Contents

The Role of Social Innovation in Climate Neutrality

Francesca Rizzo◉ and Tamami Komatsu◉

Abstract Achieving climate-neutrality by 2030 is a significant challenge for cities involved in the EU 100 Climate-Neutral and Smart Cities Mission. A systemic approach, involving cross-sector collaboration, is essential for cities to identify emission challenges, co-create actions, and implement solutions effectively. Social innovation can support this in meaningful ways and is a crucial component in achieving a just transition, offering a collaborative, inclusive, and people-centered approach to address complex challenges like climate-neutrality. It can support cities accelerate the pace of change through innovative business models, policies, and processes that mobilize knowledge, experience, and resources. This chapter emphasizes social innovation's role in three primary ways: as a collaborative process for multi-actor engagement, as platforms for action and inclusion, and as solutions to emerging needs resulting from the transition, taking evidence from the experiences thus far collected in the NetZeroCities project. It also offers directions for further research on how SI can support cities achieve climate-neutrality.

Keywords Social innovation · Climate-neutrality · System innovation · Co-creation · Engagement · Climate resilience

1 Why Cities Need Social Innovation to Achieve Climate-Neutral Futures

Achieving climate-neutrality by 2030 is an ambitious task for all cities taking part in the EU 100 Climate-Neutral and Smart Cities Mission (Cities Mission). By nature of being a mission (Mazzucato 2018), a unilateral approach to the Cities Mission is not sufficient. Instead, cities must adopt a systemic approach and act as ecosystem orchestrators to successfully define emission challenges, identify impact gaps and potential solutions, co-create a portfolio of actions and, fundamentally, co-implement

F. Rizzo (✉) · T. Komatsu
Department of Design, Politecnico di Milano, Milan, Italy
e-mail: francesca.rizzo@polimi.it

© The Author(s) 2025
S. Bresciani (ed.), *Social Innovation Projects for Climate Neutral Cities*,
PoliMI SpringerBriefs, https://doi.org/10.1007/978-3-031-87726-1_1

the actions leading to the Mission's success. The transition journey that cities embark on to achieve climate-neutral futures will require cross-sector collaboration on a massive scale and careful orchestration for simultaneous action on multiple levers of change. While technical solutions are one of the more obvious levers, social innovation (SI) stands to play a crucial role in the success of the Cities Mission in the long-term by creating the necessary impetus and buy-in for widescale change.

SI is a bottom-up, collaborative, systemic and people-centered approach to innovation. It focuses on tackling complex, wicked challenges—such as climate-neutrality—through quick, collaborative experimentation and inclusivity. SIs seek to meet unmet needs often in support of marginalized communities. SIs can be entrepreneurial, but they can also be the following: innovative business models; creative solutions that combine latent resources or new relationships and allies; policies that support new forms of value creation; or inclusive processes that mobilize a wider range of knowledge, experience and resources. In short, SIs are collaborative and mobilize communities to action through inclusion and shared value creation, and in the process, often increase society's capacity to act (Howaldt and Schwarz 2010; Hubert 2010; Murray et al. 2010). For these reasons, it is a valuable ally to a city's climate transition plan and should be a transversal component of a city's broader transition journey.

Beyond the innovations themselves, SI, as a discipline and form of innovation, has much to offer city administrators managing missions for the affinities they share in approach: both aim for systemic change and transformative impact, with the common understanding that an enterprise of such magnitude requires a collaborative approach. For this reason, it stands to offer practices and knowledge useful for accomplishing missions from a methodological standpoint as well.

SI, in practice, works best as a systemic innovation approach to generate holistic solutions to societal challenges and to create responsive ecosystems for social change. SI practices can support systems innovation by providing an inclusive and collaborative process for generating solutions that support a city's climate-neutral objectives. In doing so, SIs generate "tangible" value, seen in concrete solutions and economic development, as well as, "intangible" value, which is reflected in the potential for cultural and behaviour changes, relationship building and inclusive growth, among others.

SI can help cities accelerate their transition to climate-neutrality in many ways: (1) ensuring for consideration of economic development and overall wellbeing of people and the planet at every step of the transition to net zero; (2) highlighting the co-benefits of climate mitigation that generate social and economic value; (3) creating new business models and building capacity to address decarbonization challenges; (4) creating engagement platforms for multiple actors to co-design and co-produce solutions contributing to climate-neutrality; and (5) supporting positive behavioural changes by responding to specific local needs and acting within cultural contexts.

This chapter, and the book at large, introduce the need, role and value of SI for cities on their path to climate-neutrality, particularly within the context of the EU Cities Mission, drawing from experience coming from the first three years of the NetZeroCities (NZC) project. More specifically, the current chapter will focus on how SI contributes to a city's journey to climate-neutrality in three primary ways: (1) as a collaborative process for multi-actor engagement and collective engagement, (2) by providing platforms for action and inclusion, and (3) building solutions to existing and emergent needs resulting from the transition or climate change overall through examples coming from the NZC project.

2 NetZeroCities and EU 100 Climate-Neutral and Smart Cities Mission

Despite urgent recommendations and national agreements to curb GHG emissions and limit global warming to 1.5 °C to preserve a liveable planet, monthly and annual breaches signal the need to accelerate the pace of change and increase ambition (UN Climate Action 2024). Cities, representing 70% of global CO_2 emissions and key influencers of the way we collectively live and work, stand to play a pivotal role. For this reason, the European Union has launched the Mission *100 Climate-Neutral and Smart Cities by 2030* to support European cities in accelerating their transformation towards climate-neutrality. The Mission, along with other policies and programs—e.g. Horizon2020 and Horizon Europe research and innovation programs, the EU Green Deal, etc.—support objectives to push EU member states towards a 55% cut in emissions by 2030 and climate-neutrality by 2050 (NZC 2021). NZC is part of the Horizon 2020 program supporting the EU's Green Deal and, in practice, helps cities achieve the Cities Mission. It works as a service-oriented platform for cities, supported by a large network of expert practitioners, to develop new strategies and capabilities to enable the needed acceleration; in other words, it provides the knowledge, tools and resources to achieve *systemic* transformation. While the main goal is to support 112 Mission Cities achieve climate-neutrality by 2030, the ultimate goal is for these cities to become experimentation and innovation hubs for all EU cities to follow suit and achieve climate-neutrality by 2050.

3 The Role of Social Innovation in a City's Journey to Climate-Neutrality

SI is in a prime position to support cities in achieving climate-neutrality by 2030 or 2050 for its collaborative, inclusive and holistic innovation practices that work in tandem with a systemic approach toward transition goals. SI works on two levels: top-down support in the form of city-level strategic documents, urban planning and policy and bottom-up implementation in the form of individual solutions. We expand on this further in this section by presenting the three primary ways SI can support cities achieve the Cities Mission, as observed thus far in the NZC project.

3.1 Social Innovation as a Collaborative Process for Multi-actor Engagement and Collective Impact

The complicity between a systemic approach and SI in supporting missions has already been discussed. As such, investing in and amplifying SI as a strategic part of a city's portfolio of actions can strengthen a city's effort to achieve a just transition. Missions, as in solving most wicked challenges, require widescale collaboration, and ultimately buy-in, from all stakeholders for success. To be effective at scale will require not only innovations (social, technological, financial), but innovative processes for strategic action. SI has already built-up expertise and a repertoire of tools, methods and practices from different fields to bring a bottom-up approach to top-down strategies and policymaking (Reynolds et al. 2020). Similar to co-created policies that aspire to effectively respond to challenges, engaging diverse communities of stakeholders involved in the emission domain challenge areas through collaborative practices could generate co-created city climate action plans that are, in themselves, SIs (see also Chap. 3). The action plans—which in the NetZeroCities project are part of the Climate City Contracts (CCC),—then act as 'boundary objects' (Star and Griesemer 1989) around which diverse stakeholders gather for reasons of shared interest and negotiate and give shape to a series of commitments and strategic actions to achieve the Mission. In doing so, cities create larger buy-in by directly involving stakeholders in the city's strategic plans and increasing the success in implementation across sectors, levels and localities of the urban context. In the process of co-creating climate action plans, actors bring situated knowledge of the challenge from diverse perspectives and through inclusion ensure that the plan and its actions are directed at real problems. In other words, through the process, the actors embed the plan into the city's specific ecosystem of resources, solutions, and networks of actors. In doing so, the city increases the chance for successful implementation by creating the necessary buy-in to co-produce solutions through a shared agenda and the alignment of value propositions. This, once again, underscores the importance of cities as ecosystem orchestrators. While still too early to report on, several Mission cities have undertaken interesting, collaborative pathways to co-create their city's climate action plans.

We can also see similar mechanisms in policies that foster SI in specific emission domains. Energy Communities, for example, are a promising solution to reduce energy poverty and increase clean energy (co-)production. Through specific policy measures, these communities can generate social impacts and contribute to a just and inclusive energy transition through co-production with local actors (cities, private companies, and citizens). The City of Valencia provides an interesting example in their commitment to create "100 Energy Communities by 2030" (Giovannini 2023; Godson 2023; mPower 2022). The city has created neighborhood energy offices to act as a one-stop shop for citizens on the energy transition, concentrating on all issues related to energy poverty and energy savings, as well as providing pathways for frontrunner citizens interested in contributing to the energy transition through energy communities. Through these offices, the city helps vulnerable households implement energy savings through educational campaigns and light interventions through their energy savings kit (e.g. a timer to switch off WiFi router, TV, or water heater at night, insulation tape, etc.). The offices also help identify frontrunner citizens to partner with the city to create citizen-led energy communities through the creation of not-for-profit associations. In the first community in Castellar L'Oliveral, for example, a municipal roof was identified for PV installation and is managed by the community. The participation costs for community members are €600 per 0.50 kWp, yielding roughly €130/yr savings per household. The regional government contributes up to 50% of initial investment for the installation once the community is active and the Energy Office provides support on choosing the best energy company for the contract. Extra shares were bought by the Valencia Climate and Energy Foundation and distributed for free to three vulnerable households. In the future, the goal is to include them from the start, through the support of EU funding from the Power Up project. In sum, SI can be seen as a collaborative process that allows for bottom-up engagement in strategic directives, setting up the foundation for successful implementation and greater collective impact.

3.2 Creating Pathways for Action and Inclusion

While technical solutions are a key part of achieving climate-neutrality, successful implementation also requires mass uptake and adoption. In addition, and as already stated above, achieving the mission will require the active participation of all city-holders: public administration, but also businesses and citizens. This will require significant changes in lifestyles, perceived norms and behavior. SIs can accelerate the pace of change by creating pathways for broadscale participation and action. This is achieved in practice in many ways, namely through the sharing economy or the implementation of novel business models; innovative spaces that foster collective impact and build society's capacity to act; creative solutions to local needs; creating awareness to inform new choices and actions,; etc.

By implementing business models based on sharing or access instead of ownership, SI invites larger participation in climate goals. One more mainstreamed example of this is bike sharing programs, in which private companies often partner with municipal agencies to provide convenient micro-mobility alternatives to private transportation that make use of fossil fuels. While the assessment of potential for emissions reductions is still under scientific debate (Li et al. 2022; Chaniotakis et al. 2023; Gebhardt et al. 2022; Krauss et al. 2022), e-bike sharing and e-scooter sharing models give way to micro-mobility actions that encourage more sustainable lifestyle choices, especially when compared with situations in which the user would have used fossil fuel-powered cars.

Other cities are creating innovation labs to foster climate action at the neighborhood level that respond to local needs. Many such examples can be seen in the EU Horizon 2020 research project, SONNET, that worked to understand how SI can contribute to a more sustainable energy system in Europe. Bristol City Council, for example, created a city lab (Humphreys et al. 2021) to search for ways to make use of crowdfunding—specifically a Community Municipal Bond (CMB) mechanism—as an investment activity to collectively raise capital to install energy efficiency measures in local community buildings. In Mannheim, the city lab (Hoffman et al. 2021) aimed at mobilizing citizens through participatory stands, a gamification challenge and crowdfunding measures in a neighborhood composed of mostly migrants, where language barriers posed a challenge to the city to engage with citizens for energy transition efforts. All these efforts work to build the innovation capacity of cities by building local competences and new networks of relationships are presented more extensively in Chap. 5.

Other SIs target behavior change and lifestyle changes more directly, for example, through labels that promote consumer choice. One example of this can be seen in the Climate Meal label in Helsinki, an initiative of Forum Virium Helsinki, in which restaurant customers could identify meals from the menu that have a smaller-than-average carbon footprint (see Chap. 5). In sum, by creating enabling pathways and access points for diverse stakeholders to take part in climate action and in the transition, SI offers cities, its citizens and urban stakeholders the opportunity to make achieving the mission a truly distributed social accomplishment.

3.3 Respond to Emerging Needs

SI, also, responds agilely to the emerging needs of different communities resulting from the transition. As the transition continues and systems change, SI will respond more and more to emerging needs coming from people directly affected by climate change and/or the transition and those "inconvenienced" by the transition (i.e. those who benefitted from the challenge persisting). Because of its collaborative working practices, SI can rapidly and effectively respond to emergencies, gathering actors to co-design and co-produce solutions to the specific need. One emblematic example is the #WirVsVirus campaign that emerged in Germany to respond to the challenges coming from the COVID-19 pandemic. The initiative is an example of an open Social Innovation. It mobilized actors from across all sectors (civil society, government and the private sector) in a 48-h hackathon to develop ideas responding to the emerging challenges of the pandemic. #WirVsVirus also provided an implementation programme (130 teams for 6-months) that supported the social innovators turn the ideas into solutions. Ideas ranged from how to quickly digitalize health-care services to how to help citizens cope with lockdown-induced isolation, and how to respond to increasing instances of domestic violence. In total, twenty-eight thousand citizens with a broad spectrum of personal and professional backgrounds participated. Examples such as these demonstrate the important role that SIs play when facing climate change and other global risks. Investing in SI and other actions of citizens engagement at an early stage in the Mission is not only strategic for mass adoption of technical measures and broadscale lifestyle changes but also in terms of creating the social competences, networks and capital necessary to react urgently. As will be further explored (see Chap. 6), SI is an important tool for building climate resilient urban 'places'.

Social cooperatives are another great example of the capacity of SI to respond to emerging local needs. This legal form allows citizens to meet their own needs (economic, social or cultural) through a co-owned and democratically-controlled enterprise. Cooperatives represent a powerful vehicle for cities to adapt to climate change and increase resilience for several reasons. Cooperatives are deeply-rooted in communities, providing stability and a tried and tested solution for sustainable development. They also work on multiple objectives at the same time, adopting a long-term viewpoint, veering away from short-termism. Cooperatives also add to the resilient capacity of a city by increasing human capacity and social capital through training, member engagement, addressing local needs and developing the local economy. Cooperatives are working on the climate change across sectors from transportation to agriculture to energy to finance to retail to housing.

The EWS, ElektrizitätsWerke Schönau, for instance, started out with a group of citizens in the Black Forest who wanted to control where their energy came from. Today, they are one of the leading 100% renewable energy providers in Germany. HesbEnergie is another such example of a citizen-owned cooperative producing renewable energy for their community. Both examples not only work in energy production but in awareness raising and competence building that support other similar plants and cooperatives to emerge (presented in Chap. 5). In sum, cooperatives represent a systemic solution to help cities transition to low-carbon economies and achieve climate-neutrality. Governments would do well to create the right enabling conditions for them to flourish (e.g. legal recognition, access to markets and finance, etc.) (Borzaga et al. 2020).

3.4 R&I Agenda for Social Innovation and Climate Neutrality

In this context—where city action is both needed and increasingly taken—it is crucial to examine and understand where and how Social Innovation can support the decarbonization challenge in cities. If we examine Social Innovation from this perspective what is firstly evident is what kind of research and innovation (R&I) is needed to support cities in accelerating decarbonization efforts, and then to plan an R&I agenda accordingly. From the academic community, research and development agendas have been proposed for several aspects of Social Innovation for climate neutrality that are still in infancy phase and need further research (Howaldt et al. 2021).

Firstly, social innovation can easily create tension with policy silos and related policies, as they do not keep themselves within the boundaries of defined policy domains while developing solutions for societal problems. Many social innovations operate with this tension between traditional "top-down" policies and "bottom-up" initiatives. In this respect, awareness campaigns for policy makers are needed regarding what social innovation can contribute to decarbonization policies and how social innovation can help to reach decarbonization goals. Research can help to highlight successful examples of the interplay between decarbonization and social innovation and can assist in developing suitable governance models.

The second one refers to the need to assess how social innovation may contribute to climate neutrality. The development of a set of indicators on social innovation and climate neutrality is an essential first step for assessing the effectiveness and impact of people-centered solutions. Bresciani et al. (2024) provide a framework to evaluate progress towards carbon neutrality at the urban level, ensuring that social dimensions are integrated into environmental strategies with an extensive list of quantitative indicators and qualitative open-ened questions for progress monitoring and sensemaking, catalogued according to ten progressive phases of social innovation development at urban level.

The third aspect relates to the testing of social innovation strategies in diverse contexts. Many social innovations start on as small-scale and are very locally situated, which causes them to generally have a problem in getting attention and recognition from policy makers (Martin et al. 2015) and scaling up. Social innovations are developed in a specific local context for a specific local societal problem. Upscaling within the city or replication in other cities is, therefore, a challenge, and probably not possible for many social innovations in their complete form. Deployment of innovative business models, public–private partnerships, collaborations with businesses and public authorities, and targeted replication and upscaling strategies for the (core elements of the) social innovation can help to solve this issue. Research can support these solutions through development of tailored strategies and adequate platforms and business models for upscaling and replication, and development of appropriate forms of cooperation with local governments or businesses. Research can further give insight in how to deal with the question of whether the complete social innovation could be upscaled, and how and when this should be done.

Finally, as mentioned above, SIs flourish under the right enabling conditions, putting policymakers in the unique position to boost their emergence and impact. Research in the following areas would be timely to help policymakers understand the potential for SI and the conditions that allow it to emerge and grow: (1) rationales for action on the macro-level (EU, national and regional); (2) what SI is doing for climate-neutrality at the micro-level; and (3) systemic impact measurement tools and approaches that offer a comprehensive and detailed overview of the transition to low carbon economies and climate-neutrality.

4   Conclusion

In sum, cities stand to benefit from adopting SI as a strategic asset of their climate action plans as a means to deploy a just transition and to accelerate the pace of change through inclusion and meaningful collaboration. Cities should work to foster more strategic bottom-up social innovation practices while also developing more effective and impactful SI programming. By the latter, we mean measures that aim to support city practitioners in amplifying and scaling SI impact, that is, in supporting innovators—within the public administration but also all local stakeholders—in bringing their ideas to life through the means of SI. SIs flourish in ecosystems that provide the enabling conditions for innovation (capacity building, access to funding, access to markets, network support, etc.). Cities can support social innovators and amplify their collective impact in several ways: (1) creating and nourishing a robust ecosystem for SI; (2) acting as ecosystem orchestrators in these ecosystems; and (3) contributing to these ecosystems by removing barriers and/or filling gaps—e.g. by creating SI policies that support their growth and development, and eventually amplifying their impact through scaling/replication mechanisms.

The rest of the book dives deeper into specific areas where SI can support cities in their journey to climate-neutrality. Chapter 2 looks at how integrating SI in a city's portfolios of action can strengthen the long-term success of cities' climate ambitions. Chapter 3 identifies ways to bolster the strategic potential of SI in a city's transition journey. Chapter 4 focuses on strategies cities can use to scale successful SIs at the urban level. Chapter 5 presents SI case studies that exemplify its role in creating engagement, behavioral change and in ensuring a just transition. Chapter 6 investigates the relationship between citizen engagement and social innovation through a case study on the Wiener Klimateam. Chapter 7 presents tools and canvases for cities to strengthen SI programming and empower bottom-up SIs. Finally, Chapter 8 broadens to topic of social innovation from climate neutrality, connecting social innovation to the New European Bauhaus principles.

In conclusion, this book is part of a much broader and ambitious project and Mission that is currently supporting cities achieve climate-neutrality by 2030. It, in essence, provides a snapshot of the signals and probes that we have collected and synthesized in the first three years of the NetZeroCities project, and hopes to be a starting point for further exploration, practical implementation and learning-by-doing, as researchers, practitioners, city administrators and more tackle the pressing challenges and social changes that climate change requires of us.

References

Borzaga C, Galera G, Franchini B, Chomento S, Nogales R, Carini C (2020) Social enterprises and their ecosystems in Europe. Comparative synthesis report. European Commission. Retrieved from https://europa.eu/!Qq64ny

Bresciani S, Rizzo F, Mureddu F (2024). Assessment framework for people-centred solutions to carbon neutrality: a comprehensive list of case studies and social innovation indicators at urban level. Springer Nature Switzerland, Cham

Chaniotakis E, Straubinger A, Antoniou C (2023) Environmental impact assessment of bike-sharing considering the modal shift and lifecycle emissions. J Clean Prod 302:127008. https://doi.org/10.1016/j.jclepro.2021.127008

Gebhardt L, Kagerbauer M, Vortisch P (2022) Life cycle assessment of greenhouse gas emission reduction through bike-sharing systems. Transp Res Part d: Transp Environ 98:102979. https://doi.org/10.1016/j.trd.2021.102979

Giovannini S (2023) How to make sure energy communities thrive in your city? Retrieved from https://energy-cities.eu/how-to-make-sure-energy-communities-thrive-in-your-city/

Godson A (2023) Bringing the energy transition to people. Retrieved from https://eurocities.eu/sto ries/bringing-the-energy-transition-to-people/

Hoffman S, Reith V, Seus S, Stadler M (2021) Report on the SIE city lab in Mannheim—Sonnet Energy. Retrieved from https://sonnet-energy.eu/wp-content/uploads/2022/02/SONNET_D4_ 2_MANNHEIM.pdf

Howaldt J, Schwarz M (2010) Social innovation: concepts, research fields and international trends. IMA/ZLW & IfU. Retrieved from https://www.socialinnovationatlas.net/fileadmin/PDF/ein zeln/02_SI-Concepts-and-Understanding/02_00_SI_Concepts-and-Understanding_Howaldt-Schwarz.pdf

Howaldt J, Kaletka C, Schröder A (eds) (2021) A research agenda for social innovation. Edward Elgar Publishing

Hubert A (2010) Empowering people, driving change: social innovation in the European Union. Bureau of European Policy Advisors (BEPA), 12

Humphreys L, Jones M, Bristol Energy Network, Iskandarova M, Hielscher S (2021) Report on the SIE City Lab in Bristol—Sonnet Energy. Retrieved from https://sonnet-energy.eu/wp-content/ uploads/2022/02/SONNET_D4_4_BRISTOL.pdf

Krauss S, Bossauer P, Axhausen KW (2022) Quantifying the carbon footprint of shared electric scooters and bikes. J Transp Geogr 96:103165. https://doi.org/10.1016/j.jtrangeo.2021.103165

Li W, Wang S, Zhang X, Jia Q (2022) Mode substitution and carbon emission impacts of electric bike sharing: evidence from China. Int J Geogr Inf Sci 34(12):2451–2474. https://doi.org/10. 1080/13658816.2020.1712401

Martin CJ, Upham P, Budd L (2015) Commercial orientation in grassroots social innovation: insights from the sharing economy. Ecol Econ 118:240–251. https://doi.org/10.1016/j.ecolecon.2015. 08.001

Mazzucato M (2018) Mission-oriented research & innovation in the European Union. European Commission. Retrieved from https://ec.europa.eu/info/sites/default/files/mazzucato_rep ort_2018.pdf

mPower (2022) 2: Building energy communities. Retrieved from https://municipalpower.org/best-practice-guides/guide2/

Murray R, Caulier-Grice J, Mulgan G (2010) The open book of social innovation. National Endowment for Science, Technology and the Art: Young Foundation. June 21, 2024

NetZeroCities (2021) The NZC project and EU green deal. Horizon 2020. Retrieved from https:// www.netzerocities.eu/project/

Reynolds S, Gabriel M, Heales C (2020) D5.3: annual state of the Union Report—part 1: social innovation policy in Europe: where next? (Deliverable D5.3). European Commission

Star SL, Griesemer JR (1989) Institutional ecology, 'translations' and boundary objects: amateurs and professionals in Berkeley's Museum of Vertebrate Zoology, 1907–39. Soc Stud Sci 19(3):387–420

UN Climate Action (2024) Climate change: accelerating the pace of change. United Nations. Retrieved from https://www.un.org/en/climatechange/climate-solutions

Integrating Social Innovation in Cities' Portfolios of Action: Building Robust Ecosystems for (of) Collective Action and Social Impact

Kaisa Schmidt-Thome ⓘ **and Paul Mesarcik**

Abstract This chapter emphasizes the need for cities to integrate social innovation into their innovation portfolios to effectively address the climate crisis. The chapter identifies Enablers, Mediators, and Visionaries as essential personas for cities to nurture SI effectively. Cities acting as Enablers express their support to social innovation by providing grants, and supporting grassroots initiatives. When doing so, they need to carefully evaluate criteria that may favor commercially viable innovations over other initiatives. If cities become Mediators, they focus on creating a broader ecosystem by mapping and connecting existing initiatives. Mediators would have to note that connecting as many dots as possible does not necessarily lead to shared interests. A Mediator city would thus have to provide more specificity, pointing to issues of common relevance while respecting the diversity of approaches. Visionaries take on an ambitious role of agenda setting and mission orientation, aligning diverse actors toward a common goal, such as climate neutrality. Drawing on the concept of innovation portfolios, Visionaries facilitate interconnected actions, emphasizing the importance of addressing societal challenges beyond technological solutions. While recognizing the challenges, the chapter encourages cities to transition toward a Visionary role, leveraging their existing capabilities as Enablers and Mediators. By embracing a holistic and integrated approach, cities can foster a collaborative innovation landscape, inclusive of social innovations, ultimately contributing to meaningful and sustainable solutions for climate change mitigation.

Keywords Social innovation · Cities · Innovation ecosystem · Collective action

Slowing down climate change is becoming a cross-sectoral imperative that cannot be ignored by anyone claiming to work in the public interest. Cities are both epicenters

K. Schmidt-Thome (✉)
Demos Helsinki, Helsinki, Finland
e-mail: kaisa.schmidt-thome@demoshelsinki.fi

P. Mesarcik
Helsinki, Finland
e-mail: pmesarcik@gmail.com

© The Author(s) 2025
S. Bresciani (ed.), *Social Innovation Projects for Climate Neutral Cities*,
PoliMI SpringerBriefs, https://doi.org/10.1007/978-3-031-87726-1_2

of climate changing emissions (Lwasa and Seto 2022) and laboratories of innovation for a sustainable future (Concilio et al. 2019). As a result, cities and other local governments are emerging as pivotal players in addressing the climate crisis.

In this chapter we argue that, in order to forward the ambitious mission of climate neutrality, city governments and administrators (which we refer to as cities) will need to encourage significant social and technical innovation. We will focus particularly on social innovation (SI), and the challenges and opportunities that arise when cities try to support SI. Lastly, we will present key capabilities that cities[1] may need to create robust innovation portfolios focused on climate neutrality.

Aligning with the definition of social innovation in Chap. 1, we understand social innovation as a bottom-up, collaborative, systemic and people-centered approach to innovation. We would like to point out that the term social innovation is an umbrella term that refers to a rich variety of approaches and viewpoints. Although many formal, academic definitions exist, we are interested in the practical definition; what cities, states and funders look for when they support social innovation, and which initiatives and projects rise to the surface and inform a cultural understanding of the term. In this sense, we feel it is useful to think of social innovation as a contested terrain where multiple actors cherish and feel ownership over the concept.

It is important to highlight that in the context of energy systems, social innovation is often considered too narrowly and with a degree of instrumentalization (Wittmayer et al. 2020); it is considered valuable only as a means to an end (in this case carbon neutrality) or as a complement to technological innovation. While a certain degree of instrumentalization is necessary when talking about social innovation for a particular outcome, we believe that social innovation has transformative capacity and value in its own right. Concrete examples of social innovation for carbon neutrality include a wide array of practices and approaches such as alternative business models (including energy cooperatives), prosumerism and participatory governance approaches (see Hoppe and De Vries 2018).

While social innovation can be helpful in achieving carbon neutrality (see Chap. 1), we suggest that it should be integrated into innovation portfolios. In so doing, cities acknowledge that a plurality of approaches is needed, and help create a protected niche for social innovations to thrive. Portfolios may help to highlight how social innovations are valuable in their own right, rather than just as complements to technical innovations. Furthermore, while technical innovations often demonstrate more obvious, short term commercial value, the value of social innovations is often less tangible, harder to measure, or accrue over time. Innovation portfolios could be one way of establishing a more long-term commitment to the value of social innovation.

We will highlight the key capabilities needed to support social innovation as intertwined roles or personas that cities should consider. We refer to (selective) Enablers,

---

[1] We recognise that by referring to "cities" as singular entities, we are radically oversimplifying a complex network of administrators, politicians and sectoral service provision. We do this for the sake of readability, and make some exceptions when referring to local governments as body exercising the highest authority at local level.

(connecting) Mediators and (committed) Visionaries. This triad is based on work carried out by Demos Helsinki for and with the municipalities of the Helsinki City Region in 2022, in the context of "HEVi" (Helsinki, Espoo and Vantaa Innovations https://hevinnovations.fi/en/). This chapter builds on the report (Ahvonen et al. 2022) by connecting the framing of city personas with the abundant supporting social innovation literature.

1   Cities as Enablers

Taking on the role of an Enabler is a relatively straightforward task for cities. A city administration could start by publicly expressing a curiosity towards social innovations and by inviting individual innovators to contribute to broader city initiatives. By providing grants and resources to grassroots initiatives and small-scale projects, based on open calls, the city could support already existing innovations within their region. Publicly emphasizing the value of social innovations is a critical first step in increasing awareness among the city representatives and the general public. An Enabler city could also consider running challenge competitions that reward promising solutions, or promoting limited-time experiments which allow for conscious risk-taking and learning from failures. These might lead to preliminary market dialogues which can incentivize social entrepreneurs or other innovators to meet the city's needs. The key question to consider here is whether channeling funding and recognition towards atomic attempts can help to shape a more sustainable future for the residents. Would an Enabler be able to compile a robust portfolio of actions? Or to harness the power of the individual players in climate change mitigation?

In order to highlight the impact of an Enabler and the power of challenge competitions we have considered the case of "RATKAISU 100", an innovation competition run by the Finnish innovation fund, Sitra, in 2017. The competition spanned two years and comprised three stages. Initially, the general public identified Finland's most crucial social challenge, forming the problem statement. Then, Sitra issued an open call for teams with diverse backgrounds to propose innovative solutions. The 15 selected teams received incubation support while competing to develop the most promising social innovations (Sitra n.d.).

The two winning teams, Positive CV and Headai, received a total grant of one million euros to implement their proposals. An independent jury of seven experts selected the winners based on the effectiveness, innovation, and feasibility of their solutions. These initiatives offer a number of key benefits. Firstly, Enablers can access a wide pool of potential problem solvers through an open application process. Second, competitions incentivize participation by offering monetary prizes. Additionally, these initiatives foster multidisciplinary collaboration, which is essential when addressing complex, multifaceted issues (Hartmann et al. 2016), often referred to as "wicked problems". In this sense, challenge competitions are seen as a potential remedy to the limitations of traditional public sector research and development practices. The incumbent approaches often suffer from silos, narrow interests, and bureaucratic obstacles, which hinder the generation of innovative ideas (cf. Scott and Gong 2021). Proponents of SI competitions contend that challenge prizes offer a promising approach to overcome these issues and produce solutions that current institutional structures cannot achieve (Toivonen et al. 2018). Lastly, competitions can go a long way in providing visibility to the social innovator applicants, and almost all the Ratkaisu 100 teams reported that the competition gave them legitimacy and visibility in relevant networks (ibid.).

However these competitions are not without risks and downsides. Chief amongst the critiques are that pinning social innovators against one another in an artificially competitive environment can prevent the necessary collaboration required for social innovation to thrive—as Toivonen puts it "their competitive setup conflicts fundamentally with the principles of sharing and help-giving" (2018, p. 57). Others point to the waste of time and resources represented by the "losers" of the competition, especially considering the extremely low odds of winning, and the significant labor involved in the application process and associated events (Starr 2013). Ratkaisu 100 endeavored to reduce these downsides by embedding incubation and professional development within the competition schedule, creating benefit for all participants and not just the winners (Toivonen et al. 2018).

For social innovation, the competitive environment of innovation promotion can be a particularly challenging arena. The focus on individual solutions tends to favor novelty and hasty development, whereas social innovations often build on existing and long-term work. The word "innovation" tends to guide Enabler's attention towards the direction of technology that can often offer more specific and immediate outputs. In contrast, social innovators often focus on the word "social", referring to more profound, lasting, embedded changes. Social innovation often begins with a deep understanding of the problems themselves, which may require extensive research and engagement with affected communities. These long term and embedded approaches are often incompatible with more short-term approaches of Enablers. Celebrating specific winning solutions can also give the implicit message that selected award-winning innovations could actually solve the defined problem once and for all.

Consider the European Union Social Innovation Competition (EUSIC) as an illustrative example. Now in its tenth year, it annually grants three $50,000 prizes to projects in Europe aimed at making a positive impact on the world (EIC 2023). Interestingly, despite its name, most of the winners appear to be primarily centered around technological innovation. For instance, in 2022 (under the theme of *Affordable and Sustainable Housing Districts*), the three winners included a company specializing in solar energy and storage and another that utilized AI for waste classification. While the third winner, an organization promoting cooperative housing, clearly falls under the category of social innovation, the other two are only tangentially related.

When cities act as Enablers of such initiatives, they should critically evaluate whether the criteria for these competitions and supportive practices, which often emphasize scalability and financial sustainability, inadvertently stifle grassroots social innovation in favor of more commercially viable and technologically driven innovations. It is important to recognize that the challenges of innovation competitions that we have presented, also apply more broadly to the general role of an Enabler—in so far as enabling requires a certain degree of selectivity and atomizing. Cities should contemplate how their enabling efforts can equally foster all types of social innovation, rather than favoring the most glamorous, commercially promising and technology-heavy.

2 Cities as Mediators

Taking on the role of a Mediator requires cities to expand their definition of support beyond just the financial assistance and recognition provided by an Enabler. Instead of capitalizing on promising but disjointed solutions, a Mediator would conceive social innovations as elements of a broader ecosystem. A city organization would start with systematically mapping existing initiatives in their region (see Matti et al. 2020) and uncovering the ecosystem's constituent stakeholders and innovators. The Mediator would also design and support interfaces that allow the city to remain in touch with the innovators and vice versa. This could take the form of innovation forums, regular city supported events or innovation hubs and incubators. By building broader support systems and by forging relationships between the innovation entities, the city places their trust in the power of interconnectedness to amplify innovative initiatives. These connections could be made through communication platforms tailored to enhance open dialogue and collaborative activities (Russell and Smorodinskaya 2018). Mediators typically bring together multiple types of actors, ranging from business and academic circles to grass-root organizations, and connect them with funding, as well as providing financial resources and spaces for incubators and accelerators.

A key question is whether nurturing an ecosystem actually leads to flourishing. Would a Mediator be able to witness the emergence of a robust portfolio of actions? Can creating connections guarantee actions? How would social innovation 'cohabitate' and 'survive' in a multi-purpose ecosystem?

Proponents of these ecosystems have also referred to them as cooperation environments where the actors co-evolve (Klimas and Czakon 2022). The co-evolvement would also better justify the eco-prefix borrowed from biology in the 1990s, originally used to coin business ecosystems (Russell and Smorodinskaya 2018), but it would still not explain how co-evolvement would lead to novelty or even something socially desirable. Actually, innovation ecosystem may be a "faulty analogy to natural ecosystems" (Oh et al. 2016, p. 1) that can lead to harmful policy choices including an over-emphasis on market forces. It is this emphasis on ecosystems as the ends instead of the means that is problematic, and creates a false equivalence to the uncontroversial benefit of supporting and nurturing natural ecosystems. This can be particularly problematic for non-commercial social innovations that often depend on an individual's passion and commitment to a certain cause, and struggle to compete in an ecosystem that includes more "dominant" commercially oriented "species".

Mediators that carefully nurture their innovation ecosystems can probably expect some gains, but social innovations in particular might not thrive amongst the undirected mutualities. Connecting as many dots as possible does not necessarily lead to shared interests, and at worst, can be a waste of people's time, requiring demanding meetings and negotiations. A Mediator city would thus have to provide more specificity, pointing to issues of common relevance, which in turn requires substantial expertise and reflective facilitation. Creating nurturing ecosystems would also require regular sense-making; consistently checking the relevance of certain initiatives and prioritizing a diversity of approaches.

If a city expects that only acting as a Mediator would be rewarded with a burgeoning innovation landscape, it is likely to be disappointed. Facilitating connections and encouraging interactions without a specific mission other than generic success may lack purpose. Research from the EE-funded project SI-DRIVE (Howaldt et al. 2019) shows that more than 60% of social innovation initiatives are motivated by responding to a particular societal challenge rather than economic success of profit (Domanski and Kaletka 2018). Mediators would be wise to keep this in mind, and recognise that their support and facilitation should support a broad range of motivations. Due to the local embeddedness of social innovation (cf. Terstriep et al. 2020) cities would need to operate with a governance model that respects multiple types of innovation, with a focus on locality rather than universality. Mediators could also consider supporting or founding intermediaries—social innovation hubs, labs and transfer centers—that also nurture (ibid.) clearly non-commercial innovation.

3 Cities as Visionaries

Assuming the role of Visionary means speaking to more than just a support or coordination role. As Visionaries, cities would have to take on the responsibility of agenda setting and articulating a clear sense of purpose for innovation, including social innovation. Driving meaningful change in this way could also be called mission orientation—an approach that focuses on addressing specific societal challenges by directing resources, efforts and policy towards well-defined, ambitious goals (Mazzucato 2018). A Visionary would need to take a wise and reflective approach, respecting the multiplicity of competencies and diverse motivations within its region and aiming to align actors to accelerating a desired change, such as the mission of carbon neutrality. Much of the alignment could happen through innovation portfolios[2]; collections of diverse initiatives that aim for a certain impact, enhancing mutual learning between the projects. Social innovations are likely to find their place in these portfolios, because mission orientation—when the Visionary city gets it

---

[2] By using the term innovation portfolios we refer to thought and literature related to public sector and mission-oriented innovation as opposed to the literature in "investment" innovation portfolios (cf. Martinsuo and Anttila 2022; Morris 2010; Si et al. 2022) which seek to improve private sector growth and competitiveness.

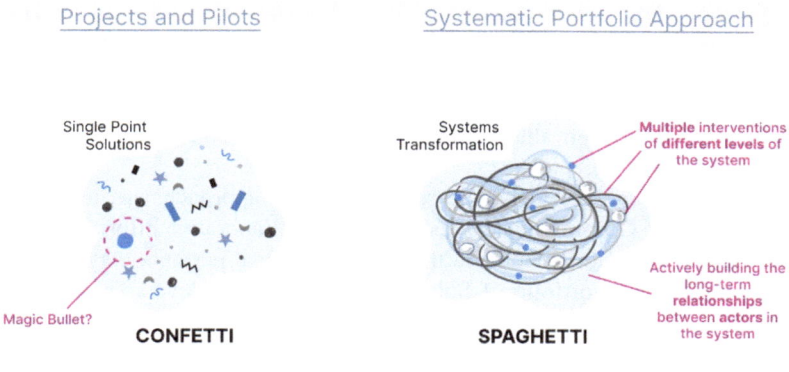

Projects and Pilots   **Systematic Portfolio Approach**

Single Point Solutions   Systems Transformation   **Multiple** interventions of **different levels** of the system

Magic Bullet?   **Actively building the long-term relationships** between **actors** in the system

CONFETTI   **SPAGHETTI**

Adapted from: Ingrid Burkett/TACSI, Sitra

Fig. 1 The confetti and spaghetti metaphor for project-by-project mentality

right—should involve more than just technological innovation alone (Foray et al. 2012).

As Seppälä (2021, p. 12) summarizes, "the portfolio approach relies on close interaction between connected projects or experiments, sensemaking between those involved in them and the shared impact arising from actions driven by that shared understanding". Utilizing metaphors, he distinguishes between the project-by-project mentality (typical of an Enabler and Mediator) that resembles a cloud of confetti—small pieces of paper creating special effects—from innovation portfolios that should rather look like spaghetti (Fig. 1), where elements are intertwined. Seppälä (2021, p. 17) recommends four sets of practices: (1) identifying challenges that require systems transformation; (2) setting out how the whole portfolio seeks to make an impact; (3) composing a collection of projects with shared intent; and (4) engaging in sensemaking, generating insights and changing activities."

However, adopting the portfolio approach, aiming at systemic changes and developing matching competences requires considerable collective efforts which the city would need to facilitate. Finding common ground around a meaningful goal or a shared mission is by no means a trivial task. The key question is whether steering of this kind is feasible for a city, and desirable for the involved actors.

Let's examine the case of the UNDP's Istanbul Regional Hub, which has been working on the Agorà model, described as a "City Transformation Portfolio." This model serves as a blueprint for intricate urban system design, encompassing urban renewal, transformation, and pandemic response (Järvelä and Chung 2021). The Agorà model is a systematic, step-by-step process[3] which assists country offices or cities in approaching complex problems while also considering how their interventions fit into urban transformation portfolios.

---

[3] Agorà was developed in collaboration with the Chora foundation.

This approach is not confined to the Istanbul Regional Hub and has been echoed in other aspects of UNDP's work, such as the Social Innovation Platforms (SIP). The SIP approach,[4] piloted in Thailand, Pakistan, and Indonesia, closely resembles innovation portfolios. It focuses on strengthening local development planning processes through a systems approach that is inclusive, participatory, and resilient. It leverages real-time data, human-centric design thinking, and a range of solutions to address complex challenges (UNDP n.d.).

Taking a closer look at a specific region, the SIP approach was implemented in the provinces of Gorontalo and West Java, in a collaborative effort between the UNDP's Bangkok Regional Hub and the ALC. Technical partners played a pivotal role in developing a "Portfolio of Solutions" at the village level. These solutions emerged from "deep listening" and co-creation processes involving various stake-holders, systematically addressing issues across five interconnected levels: community relations, small-scale businesses, large-scale public–private partnerships, public service redesign, and new regulations (UNDP 2022).

While the best examples of creating impactful, robust innovation portfolios may not stem primarily from the city organizations, there are good reasons to remain hopeful of their applicability in a more local context. In a number of countries, local authorities possess extensive administrative and legal powers and substantial financial resources derived from tax revenue. They operate in many sectors that are central to society's well-being and carbon footprint. However, local governments have not originally been established to address global challenges or to gather the collective intelligence for solving them. Becoming host to targeted, results-driven innovation policy and fostering a collective sense of purpose might sound far too challenging. It might also be hard to find political backing for such a role, not to mention the resourcing it requires. However, cities have good reason to attempt to transition to a more Visionary role even if it means that new skills must be acquired.

Many cities have already been active as Enablers (Hartmann et al. 2016; Pihla-jamaa and Merisalo 2021) or as Mediators (Robaeyst et al. 2021), gaining valuable experience, infrastructure and know-how that can be built upon. In fact, being a successful Visionary requires cities to build upon the key capabilities of an Enabler and a Mediator.

The good news is that social innovations complement the Visionary approach. With this approach, social innovators are not pushed to compete, but invited to contribute, alongside the technology based and market driven contributions. Here, the mingling of social innovators with other members of the innovation ecosystem is likely to have a purpose as the alignment of actions into portfolios is formed around an ambitious societal goal. As a Visionary, a city would care about the inclusivity of both the process and the solutions, helping to extend the benefits to marginalized or vulnerable populations. Furthermore, social innovations and mission-orientation both favor adaptive and iterative problem-solving approaches. A Visionary would

---

[4] It was developed in partnership with the Basque Social Innovation Lab—Agirre Lehendakaria Center (ALC)

acknowledge that addressing complex societal issues requires continuous learning, adjustments and experimentation.

A Visionary would have to be both ambitious in goal setting and humble while searching for solutions (see Annala et al. 2021). A Visionary would have to admit to not having all the answers itself, instead facilitating an iterative process of distributed problem solving, experimentation and continuous sensemaking among various stakeholders. Furthermore, a Visionary would need to ensure that their portfolios include a selection of public policy instruments (including regulations and taxation) both to guide and incentivise the innovators in their midst, and also to buttress against the negative externalities generated by the private innovation sector (Foray et al. 2012). Such an approach is well suited to address climate change, which is a complex and uncertain social and policy issue necessitating humility and reflexivity. A Visionary would welcome support from organizations or centers of expertise specialized in social innovation and could commission further helping hands from intermediaries that can operate at the (social) innovation interface.

4 Conclusion

We have outlined three personas that cities could consider when they support innovation more broadly, and social innovation more specifically. While the Enabler can provide quick wins and excitement, there is a risk that it leads to atomic and disconnected actions. Instead, cities could consider acting as a Mediator, and focus on connecting different actors and facilitating collaboration. Again, this is a positive and supportive position, but without giving directionality to the facilitated connections, the ecosystem is unlikely to lead to distinct outcomes, such as climate neutrality. The role of the Visionary, which draws inspiration from work on innovation portfolios (see UNDP 2022), describes a more holistic and integrated approach. In this role a city would help create meaning and groundswell to direct innovation initiatives, facilitate compatible policy, and also protect those innovations which are less obviously commercially viable. However, this role may include resources and capacities beyond the scope of cities. We see anticipatory signs that more attention and funding is being directed towards innovation portfolios, and cities may be able to harness these resources to build new capabilities. Prominent initiatives of this kind include Climate-KIC and Net-Zero Cities. Similarly, on the local level, cities are not alone. If they turn to their changing landscape of innovation and start to gather mission-oriented portfolios of action, they may soon find themselves surrounded by a crowd of aligned actors working together for the public interest.

Assuming that acting as an Enabler (only) would lead to blossoming innovation eco-systems that can cherish social innovations on equal footing with other novel approaches is most likely a false bet. By giving funds and providing spaces, guidance or other public support, an Enabler city might be able to put a set of individual innovations, including social ones, in the spotlight. While these might attract public attention, the city would hardly be able to build robust portfolios of interconnected actions. Furthermore, if an Enabler wanted to compile a set of innovators to address climate change mitigation, it could be successful in case the (often commercial) interest of the solution providers happened to coincide with that aim—or if the incentives were strong enough to attract fully new providers/innovators to join forces with the city for the sake of the climate.

If a city expects that acting as a Mediator (only) would be rewarded with mushrooming innovation landscape, it is also likely to get disappointed. Facilitating connections and encouraging interactions without a major content other than generic success is not very compelling. In particular, complementing the more technology driven solutions with social innovations is unlikely to happen if there is no sense of a direction. Due to the local embeddedness of social innovation (cf. Terstriep et al. 2020) cities would need to operate with a governance model that respects multiple types of innovation. They could also consider supporting or founding intermediaries—social innovation hubs, labs and transfer centers—that nurture also clearly non-commercial innovation (ibid.).

When a city takes the role of a Visionary it adopts the portfolio approach and gathers the various competences accordingly around the desired systemic change, such as climate change mitigation. The city admits that even in its own policy development many important insights come from outside the City Hall. It is thus worth it to open up the problematic and to facilitate the portfolio development with interested members of the innovation ecosystem. The capabilities required in this kind of orchestration have traditionally not belonged to key capabilities of city administrators or sectoral experts, but now the time would be right to change the priority order. The Visionary cities would develop these capabilities, to align a diverse set of efforts including their own, in order to thrive together with the committed set of members from the innovation ecosystem.

References

Ahvonen K, Koste O-W, Outi K, Sokero M, Davidson E (2022) Director, mediator, enabler. Analysis of the ecosystem agreement environment in the metropolitan area and the roles of cities as innovation actors in the 2020s. Demos Helsinki

Annala M, Gronchi I, Leppänen J, Mertsola S, Sabel C (2021) A call for humble governments. Demos Helsinki. https://demoshelsinki.fi/wp-content/uploads/2021/09/Humble-Government_Sept2021-.pdf

Concilio G, Li C, Rausell P, Tosoni I (2019) Cities as enablers of innovation. In: Concilio G, Tosoni I (Eds) Innovation capacity and the city: the enabling role of design. Springer International Publishing, pp 43–60. https://doi.org/10.1007/978-3-030-00123-0_3

Domanski D, Kaletka C (2018) Social innovation ecosystems. In: Howaldt J, Kaletka C (eds) Atlas of social innovation: new practices for a better future. Sozialforschungsstelle, TU Dortmund University, Dortmund

EIC (2023) The history of the European Social Innovation Competition. History of the Social Innovation Competition. https://eic.ec.europa.eu/eic-prizes/european-social-innovation-competition/history-european-social-innovation-competition_en

Foray D, Mowery DC, Nelson RR (2012) Public R&D and social challenges: what lessons from mission R&D programs? Res Policy 41(10):1697–1702. https://doi.org/10.1016/j.respol.2012. 07.011

Hartmann S, Mainka A, Stock WG (2016) Opportunities and challenges for civic engagement: a global investigation of innovation competitions. Int J Knowl Soc Res 7(3):1–15. https://doi.org/ 10.4018/IJKSR.2016070101

Hoppe T, De Vries G (2018) Social innovation and the energy transition. Sustainability 11(1):141. https://doi.org/10.3390/su11010141

Howaldt J, Kaletka C, Schröder A, Zirngiebl M (2019) Atlas of social innovation. New practices for a better future. Sozialforschungsstelle, TU Dortmund University, Dortmund

Järvelä E, Chung Y (2021) Agorà — a transformational dance with urban systems. UNDP Eurasia. https://undpeurasia.medium.com/agor%C3%A0-a-transformational-dance-with-urban-systems-197c2b1b55e0

Klimas P, Czakon W (2022) Species in the wild: a typology of innovation ecosystems. RMS 16(1):249–282. https://doi.org/10.1007/s11846-020-00439-4

Lwasa S, Seto KC (2022) Urban systems and other settlements. In: IPCC (ed) Climate change 2022—mitigation of climate change, 1st edn. Cambridge University Press, pp 861–952. https:// doi.org/10.1017/9781009157926.010

Martinsuo M, Anttila R (2022) Practices of strategic alignment in and between innovation project portfolios. Proj Leadersh Soc 3:100066. https://doi.org/10.1016/j.plas.2022.100066

Matti C, Corvillo JMM, Lalinde IV, Agulló BJ, Stamate E, Avella G, Bauer A (2020) Challenge-led system mapping: a knowledge management approach. In: Matti C (ed) EIT Climate KIC. https://transitionshub.climate-kic.org/publications/challenge-led-system-mapping-a-knowledge-management-approach/

Mazzucato M (2018) Mission-oriented innovation policies: challenges and opportunities. Ind Corp Chang 27(5):803–815. https://doi.org/10.1093/icc/dty034

Morris L (2010) Managing innovation portfolios. InnovationLabs LLC

Oh D-S, Phillips F, Park S, Lee E (2016) Innovation ecosystems: a critical examination. Technovation 54:1–6. https://doi.org/10.1016/j.technovation.2016.02.004

Pihlajamaa M, Merisalo M (2021) Organizing innovation contests for public procurement of innovation—a case study of smart city hackathons in Tampere, Finland. Eur Plan Stud 29(10):1906–1924. https://doi.org/10.1080/09654313.2021.1894097

Robaeyst B, Baccarne B, Duthoo W, Schuurman D (2021) The city as an experimental environment: the identification, selection, and activation of distributed knowledge in regional open innovation ecosystems. Sustainability 13(12), Article 12. https://doi.org/10.3390/su13126954

Russell MG, Smorodinskaya NV (2018) Leveraging complexity for ecosystemic innovation. Technol Forecast Soc Change 136(C):114–131

Scott I, Gong T (2021) Coordinating government silos: challenges and opportunities. Glob Public Policy Governance 1(1):20–38. https://doi.org/10.1007/s43508-021-00004-z

Seppälä M (2021) Radical uncertainty requires radical collaboration. Sitra

Si H, Kavadias S, Loch C (2022) Managing innovation portfolios: from project selection to portfolio design. https://ssrn.com/abstract=4050940

Sitra (n.d.) Ratkaisu 100. Sitra. Retrieved 12 Oct 2023, from https://www.sitra.fi/en/topics/ratkaisu100/

Starr K (2013) Dump the prizes (SSIR). Stanford Soc Innov Rev. https://ssir.org/articles/entry/dump_the_prizes

Terstriep J, Rehfeld D, Kleverbeck M (2020) Favourable social innovation ecosystem(s)?—An explorative approach. Eur Plan Stud 28(5):881–905. https://doi.org/10.1080/09654313.2019. 1708868

Toivonen T, Nordbäck E, Takala V (2018) Sparking social innovation through challenge prizes (133). Sitra Studies

UNDP (n.d.) Social Innovation Platform (SIP). https://www.undp.org/asia-pacific/social-innovation-platforms/social-innovation-platform

UNDP (2022) System change: a guidebook for adopting portfolio approaches. United Nations Development Programme Bangkok Regional Hub

Wittmayer JM, De Geus T, Pel B, Avelino F, Hielscher S, Hoppe T, Mühlemeier S, Stasik A, Oxenaar S, Rogge KS, Visser V, Marín-González E, Ooms M, Buitelaar S, Foulds C, Petrick K, Klarwein S, Krupnik S, De Vries G, Härtwig A (2020) Beyond instrumentalism: broadening the understanding of social innovation in socio-technical energy systems. Energy Res Soc Sci 70:101689. https://doi.org/10.1016/j.erss.2020.101689

Strategizing for Climate Neutrality. How to Make Social Innovation Relevant in the City's Journey to Climate Neutrality

Marzia Mortati and **Eugénie Cartron**

Abstract This chapter explores the imperative for cities to accelerate their journey towards climate neutrality through social innovation and systemic transformation. It begins by contextualizing the challenge within the European Union's "100 Climate-Neutral and Smart Cities by 2030" Mission, emphasizing the need for collaborative governance and creative approaches to city planning. Key practices for sustainability action planning are outlined, including the establishment of Transition Teams, development of portfolios of actions, mobilisation of cross-sectoral ecosystems, and iterative planning processes. The chapter further introduces the Climate Transition Map as a tool to guide cities through their transition journey. Additionally, it discusses the importance of learning and capability building for embedding sustainability and social innovation in city practices, highlighting the need for facilitating continuous learning and reflection for informing innovation towards sustainable practices. Finally, it addresses the evolving nature of sustainable transitions and the need for transformative policies, innovative approaches, and agile structures to achieve systemic transformation. Throughout, the chapter emphasizes the central role of social innovation in empowering collective action towards climate neutrality, as a crucial aspect of sustainability transitions beyond technological innovations.

Keywords Social innovation · Sustainable transitions · Transition team · Systemic innovation · Transformative policy

M. Mortati (✉)
Department of Design, Politecnico di Milano, Milan, Italy
e-mail: marzia.mortati@polimi.it

E. Cartron
DarkMatterLabs, London, England

© The Author(s) 2025
S. Bresciani (ed.), *Social Innovation Projects for Climate Neutral Cities*,
PoliMI SpringerBriefs, https://doi.org/10.1007/978-3-031-87726-1_3

1 Introduction

The European Commission has made a strong statement through the European Green Deal and the 100 Climate-neutral and Smart Cities Mission: any climate transition must also be a just transition. This underscores the crucial role of social innovation in accelerating transformation processes, where conventional policy solutions are no longer effective in addressing global challenges. To tackle problems like inequality, climate change, and migration, new methods, partnerships, and ways of thinking are needed, allowing for the broader participation of people and society while directing innovation toward more expansive social purposes.

Social innovations involve "new ideas (products, services and processes) which simultaneously satisfy social needs more efficiently than existing ones and create new and long-lasting social relationships and collaborations. Not only are these innovations good for society, they also improve its ability to act" (Hubert 2010). These innovations thrive on experimentation and prototyping, allowing for the rapid identification of limitations and potentials. They also involve a creative and collaborative process to synthesize innovative concepts that address societal needs holistically, including human, social, ecological, and economic aspects. Social innovation hinges on stakeholder participation, leveraging their experience and creativity to tackle challenges. It facilitates change by shifting participants' perceptions and enhancing their willingness and ability to act (Grimm et al. 2013).

Although central in the journey towards climate neutrality, social innovation is hard to be adopted and implemented in city public administrations. Barriers include a lack of widespread acceptance and the difficulties of changing entrenched bureaucratic procedures that hinder innovation. Overcoming these internal barriers is a daunting task, often leading public officials and civil servants to abandon innovative approaches. Consequently, the results often demonstrate only partial achievements of climate neutrality goals, focusing primarily on technical and quantitative aspects like reducing greenhouse gas emissions or exploring new technologies for specific applications. While these aspects are crucial, they offer only partial solutions, potentially overlooking the broader systemic and interconnected elements crucial for steering society in a new direction. These include sustaining behavioural change and adopting new social norms. Achieving climate neutrality demands collective efforts to overcome these obstacles. It requires promoting economically viable yet socially just innovations while also institutionalizing these new practices and adopting new working principles inside public administrations (e.g., co-creation). This holistic approach is essential to address the intricate challenges associated with climate neutrality comprehensively and effectively.

This chapter offers insights and practical advice to public officials and civil servants seeking to integrate or enhance social innovation practices within their organizations and local contexts, fostering sustainable transitions. It proposes two approaches to adopting social innovation for sustainable transitions. Firstly, by positioning social innovation as an outcome of innovation policies, supporting initiatives in the social economy, investing in innovation skills, and cultivating a robust market for social innovations. Secondly, by embracing social innovation principles as an approach to address policy challenges. This involves engaging more people in policy design and collaborating with civil society organizations in innovative ways, such as co-creation processes, to tackle social issues effectively. More specifically, the chapter explores why sustainable transitions demand social innovation and introduces the NZC Climate Transition Map (CTM). This tool was developed as part of the NetZeroCities project (https://netzerocities.eu/) to identify the main phases a city goes through when approaching the complex journey to climate neutrality. The CTM outlines seven essential phases for an accelerated transition: a strong mandate, system understanding, co-created portfolio, action, learning and reflection, institutionalization, and activation of an inclusive ecosystem for change. The chapter underscores the integral connection between these phases and social innovation, particularly as a strategy for crafting socially innovative public policies essential to advancing climate transitions. Social innovation plays a pivotal role by enhancing the local ecosystem's capacity to drive bottom-up change, thereby bolstering the mandate for transformation through collaborative efforts. Furthermore, it provides valuable methods, tools, and heuristics for comprehending complex systems, facilitates the co-creation of portfolio solutions, enables collaborative action and learning through experimentation, and normalizes novel solutions by emphasizing their scalable impact.

Finally, this chapter underscores the introduction of social innovation within organizations through the lens of collective learning. Social innovation, as elucidated by Nilsson (2023) is a collective capability rather than an individual trait. Due to its foundation in co-creation, impactful social innovation cannot thrive in isolation but requires collective engagement. Consequently, coaching individuals in this area proves significantly less effective than coaching entire teams, networks, or even whole organizations. Individual training, when reintegrated into a collective context, might result in unsustainable individual change. Therefore, central to the approach detailed in this chapter is the introduction of the concept of the Transition Team. This concept becomes pivotal for cities aiming to cultivate essential cross-sectoral collaborations and acquire new capabilities. Unlike traditional organizational structures, a Transition Team transcends conventional boundaries, both organizational and sectoral. It fosters collective action and alignment among local stakeholders, which are imperative for transformative change. Serving as the orchestrator of local actors engaged in transformative endeavours, the Transition Team acts as a steward of the local ecosystem. It efficiently absorbs and disseminates novel ideas, practices, and mindsets. Crucially, the Transition Team assumes a vital role in the acquisition and dissemination of capabilities essential for achieving climate neutrality. In essence, it forms the linchpin for the efficient integration of social innovation into the fabric of organizational operations.

2   Sustainable Transitions Require Social Innovation to Address the Root Causes

The global call for new governance approaches and innovations to accelerate sustainability transitions has resonated across influential organizations such as the United Nations, the European Environmental Agency, the OECD, the German Advisory Council on Global Change, the European Union, and numerous national governments. These entities have released reports and policies advocating for sustainability transitions, highlighting the imperative of "transformative change." This urgency stems from the pressing need to confront critical global challenges such as climate change, biodiversity loss, resource depletion, and inequality. The shared objective is to achieve a just and sustainable future within ecological boundaries by 2050. Despite ambitious commitments, significant investments in innovation, and voluntary initiatives, these challenges persist, and our economies persist in unsustainable trajectories, continuously pushing ecological limits. As a result, the emphasis on transformative change and innovative governance approaches has intensified, aiming to reshape our societies towards a sustainable, equitable, and ecologically balanced future.

The challenge of changing established courses, especially through controlled or incremental strategies, has been a subject of extensive research from various disciplinary angles. The socio-institutional perspective in sustainability transitions research (Grin et al. 2010; Loorbach et al. 2017; Markard et al. 2012) illuminates how policies and innovation efforts often fixate on optimizing existing systems, leading to entrenched patterns and resistance against change. Transition governance and management (Loorbach 2010) point out that policy-driven innovation tends to assume linear, planned trajectories for technological advancements that will eventually scale and diffuse through market mechanisms. This approach may neglect other forms of innovation that emerge more organically in society where technology plays a less dominant role, such as new lifestyles, business models, or organizational structures. Such innovations are deeply influenced by local contexts and political cultures, and driven by engaged citizens or entrepreneurs responding to unique challenges and opportunities. Further, these local initiatives are not isolated occurrences; they are part of extensive networks spanning regions, where ideas, objects, and activities are exchanged and adapted. Together, they craft a "transformative innovation," marked by shared discourse, objects, and practices. This collaborative process involves intricate social and material elements, forming a complex, evolving, and multi-participant phenomenon (Avelino and Wittmayer 2016). These innovations, deeply rooted in their geographic and cultural contexts, are pivotal in achieving sustainable transitions, complementing technological advancements. Exploring and amplifying these transformative innovations can unveil novel sources and mechanisms for sustainable transitions, offering public administrations fresh perspectives and avenues for innovative practices.

The experimental efforts associated with social innovation can yield radically alternative solutions to unsustainable economic models. These endeavours, often local and tangible, empower citizens, mobilize resources, and foster adaptation through small-scale actions. Recognizing this potential, the discourse on socio-technical innovations for sustainability transitions should be expanded, delving into how locally implemented shared ideas and activities address the root causes of unsustainable behaviours. This exploration should extend beyond technology, encompassing socio-economic factors such as institutional injustices, exploitation, and inequality. For instance, new capabilities and organizational structures should be explored to understand how public administrations might concretize opportunities and raise awareness for sustainable transition goals. This might also include technology but often focuses on providing for basic needs (e.g., food, energy, water, mobility, housing), aiming at broader changes in the social fabric. Considering for instance the transition toward a circular economy, this involves more than just adopting new materials and production methods. It encompasses behavioural shifts like product sharing, novel concepts such as "waste-free living," and innovative business models where companies offer services rather than products, assuming responsibility throughout the entire lifecycle. Therefore, achieving sustainability transitions necessitates not only technological innovations but also social innovations. In other words, tackling the root causes requires the implementation of novel ways of thinking, doing, and organizing. This demands collaborative efforts between expert teams and local knowledge, working together toward a shared goal.

2.1 Key Areas in the Link Between Social Innovation and Sustainable Transitions

Many initiatives targeting greenhouse gas (GHG) emissions bring benefits beyond climate change mitigation, notably reducing air pollution and its associated health and environmental risks. These positive outcomes also include resource efficiency, economic stability, and ecosystem sustainability (UNECE 2016). City and regional governments are ideal platforms to integrate these benefits into decision-making, as interventions can yield immediate impact (Jennings et al. 2020). Increased investments in climate-related endeavours can boost economic growth and employment, yet they might also disrupt economic patterns, leading to winners and losers (UNECE 2016). Economist Thomas Piketty (2022) highlighted that economic growth alone does not guarantee improved income distribution and can worsen inequalities. Social innovation can offer a way to balance inequalities by prompting governments to reform policies, increase public participation and policy ownership, improving the effectiveness and efficiency of public services. To capitalize on these opportunities, governments are recognizing the importance of policy reforms with the aim to facilitate social innovation without suppressing its autonomy and spontaneity (Grimm et al. 2013).

Despite positive signals, however, the link between sustainable transitions and social innovation is not yet widely understood and accepted. A few key areas of concern still need to be addressed as outlined below.

Social aspects. Social innovation is closely linked to the renewed focus on the social dimensions of sustainability transitions. Ecological sustainability is deeply entwined with the desire for a more democratic, inclusive, and just society. This isn't merely a matter of individual beliefs; it's a response to the increasing demands of present and future generations, including the upcoming workforce. At a time marked by heightened attention to problems such as racism, gender disparities, abuse, and multiple forms of exclusion and inequality, no organization can effectively tackle sustainability without addressing these issues.

Collective strength. Exploring the catalysts behind social innovation reveals a diverse landscape of influencers. While governments, major market players, and educational institutions wield significant impact, a vibrant realm of collective action thrives within communities and civil society. Take, for instance, the resurgence of cooperative enterprises, particularly at the local level, where civic initiatives and collective efforts are gaining momentum. A prime example is community-based energy generation cooperatives, where consumers also assume the role of producers. Notably, numerous social innovations originate from the civil society, giving rise to innovative organisational models that serve as alternatives or hybrids, melding elements of community, market, and government. Consider collaborative ownership structures like heat grids involving public, private, and cooperative entities, or the growing presence of social entrepreneurship and socially oriented investments blending public welfare with business principles. By rebalancing power dynamics, social innovation holds the potential to steer towards valid alternatives to the currently dominant socio-economic systems.

Businesses and governments mindset. To empower community-based social innovation, a fundamental shift in the attitudes of businesses and governments is imperative. Historically, governments have heavily invested in nurturing entrepreneurial environments through initiatives such as grants, incubators, start-up support programs, innovation incentives, and collaborative networking events designed to foster innovation before competition sets in. Imagine if similar substantial resources and efforts were dedicated to fostering innovation emerging organically from within communities. Many initiatives have shown promise, but they often struggle to not only survive but also expand and scale. Community-driven innovation consistently faces challenges, including financial constraints, the need for institutional support and space, heavy reliance on volunteers, and sometimes a lack of recognition, despite their remarkable achievements. Businesses need to undergo a transformation as well, and there are global examples demonstrating how companies can closely collaborate with communities. Innovative corporate models like Fairphone serve as inspiring examples. They engage a user platform for both technical support and repairs, creating a mutually beneficial relationship. In essence, there are numerous avenues for businesses to effectively harness the collective power of communities in the pursuit of sustainable and equitable transitions.

Power relationships. While social innovation is often hailed as a force for positive change, it is essential to approach it with a discerning eye. At its core, social innovation involves reshaping relationships among diverse actors: consumers, producers, retailers, citizens, administrators, and businesses. These shifts in social dynamics inherently trigger changes in power structures. During significant societal transformations, these power relations inevitably undergo shifts, giving rise to new forms of imbalances, exclusion, and potential misuse. Consider the platform economy, where novel revenue models, exemplified by companies like Uber or Airbnb, can lead to concentrated power and pose challenges to local and regional economies, both socially and environmentally. Governments are becoming increasingly aware of these complexities. Many local authorities are now proactively regulating the platform economy, emphasizing the importance of upholding public values in the face of these shifts. This example underscores the need for a nuanced and informed perspective on the consequences of altering power balances, whether these changes originate top-down or bottom-up. Social innovation, while promising, is a realm where careful consideration and a balanced approach are crucial to navigating the intricate web of power dynamics.

2.2 Social Innovation Coupled with Systems Innovation

To drive sustainable transitions effectively, social innovation must be coupled with a robust systems innovation strategy. This combination generates comprehensive solutions for societal challenges and establishes adaptable environments for catalysing social change. Social innovation practices contribute significantly to systems innovation by fostering an inclusive and collaborative approach to problem-solving, aligning with a city's objectives for climate sustainability. In this intricate process, social innovations yield both tangible and intangible benefits. Tangible gains include concrete solutions and economic progress, while intangible advantages encompass potential cultural shifts, enhanced relationships, and inclusive economic growth, among others.

Within the public sector, social innovation plays a pivotal role in achieving climate neutrality from two key perspectives. Firstly, social innovations can be policy outputs, that is, policies for social innovation may result in creating conducive conditions for a flourishing social market economy, encompassing social entrepreneurship and enterprises. Secondly, social innovation can be an approach to policy making, that is, principles of co-creation and experimentation can be integrated into the formulation and implementation of new public policies. These policies are challenge-focused, rooted in co-creation and experimentation, and capable of nurturing new skills across the public sector (Reynolds et al. 2016). Under this framework, a diverse array of stakeholders, including individuals, civil society, and service providers, actively participate in the policy-making process through bottom-up or participatory techniques. Their involvement helps shape strategy and priorities in response to policy problems. By harnessing localized expertise and insights into the needs and aspirations of people, social innovators, and other contributors, these techniques empower policymakers to better understand and tailor policy responses. Furthermore, citizen participation, especially that of "passive citizens," can enhance confidence in the political system, deepen understanding of political objectives, and improve governance and decision-making. Overall, these participatory and collaborative approaches not only mitigate political risks associated with top-down initiatives but also significantly enhance their likelihood of success.

However, implementing these changes within cities is a highly intricate task. Cities vary significantly in size, population, density, economic activity, and culture. Achieving substantial emissions reduction within cities necessitates a deep understanding of their individual systems. Crucially, this understanding should prioritize the integration of climate initiatives throughout the entire city, cutting across diverse sectors, rather than concentrating solely on municipal endeavours. Effective collaboration with a wide spectrum of stakeholders at both local and broader governance levels is indispensable.

To drive such transformative changes, rapid integration of adjustments into local government structures is essential. This includes establishing new forms of collaborative governance and promoting flexible learning and innovation practices, often facilitated by what we term "Transition Teams." Cities must take the lead in orchestrating efforts to realise this collaborative transition, efficiently aligning and coordinating the actions of local stakeholders across various sectors. To achieve this, a diverse range of skills and competences is indispensable, including transition leadership, ecosystem collaboration, and a variety of practical, data-oriented abilities. At its core, this transformation requires a fundamental re-evaluation of the culture and role of local government. Enabling this transformative work differs significantly from the conventional service provision functions typically associated with public administration.

Adopting this perspective, we propose in the remainder of the chapter a way to join a social innovation perspective with a systems innovation approach to accompany cities in their transition, a process that we have called the Climate Transition Map.

3 Sustainability Action Planning as a Systemic and Social Innovation Journey: The Climate Transition Map

Committed to leading the charge towards continental climate-neutrality, the European Union has launched the "100 Climate-Neutral and Smart Cities by 2030" Mission, also known as the "Cities Mission". This mission is part of a broader mission-oriented policy initiative aimed at tackling pressing challenges like climate resilience, cancer, ocean restoration, and soil care by leveraging expertise, public engagement, and collaborative efforts. The Cities Mission, supported by NetZeroCities (NZC), its Mission platform, underscores the imperative for cities to build new capabilities to accelerate their journey toward climate neutrality with and for citizens. Further, it reinforces the demand for new and creative approaches to governance, public involvement and city planning, leveraging social innovation to create an inclusive pathway to a sustainable future. This approach involves testing and demonstrating changes while ensuring active participation and shared confidence in the future being built (European Commission 2021). In other words, achieving climate-neutrality requires a society to act differently, understanding emergent and unmet social needs collaboratively to unlock the technical, behavioural, institutional barriers hindering change.

Change necessitates adjustments in organizational structures, practices, and cultures within local governments. This shift involves several key long-term transformations in public administrations that we explore in the following paragraphs, namely: (1) the adoption of new organizational structures like Transition Teams, tasked with fostering extensive local collaboration, leveraging social innovation processes; (2) the development of new ways for planning around portfolios, gathering actions across social, economic and technical systems; (3) the practice of mobilising a cross-sectoral ecosystem to share the efforts of climate transitions; (4) the development of iterative processes to understand the way forward, realizing that such transformations act simultaneously at the macro and micro levels, touching cities as entrenched systems and individuals in their daily environments.

Traditional municipality-centred planning processes are to evolve toward inclusive collaborative ecosystems and iterative, learning-oriented attitudes. Local governments will need to move away from traditional roadmaps, instead co-developing dynamic portfolios combining technical and social forms of innovation with local stakeholders. The following sections introduce further these key practices NZC aims to build with local governments of the 100 Mission Cities to accelerate transformative change.

3.1 The Transition Team

Addressing the challenge of achieving climate neutrality by 2030 requires a departure from traditional management methods and hierarchical structures. Instead, it calls for a transformative approach to organizing and coordinating efforts within a city. This transformation necessitates unprecedented collaboration and coordinated action from diverse stakeholders. In the Cities Mission, the concept of the Transition Team is a dynamic and adaptive response to the complex and interconnected nature of impactful climate action. In essence, a Transition Team functions as a forward-looking partner within the local government, playing a pivotal role in guiding the city and local communities toward a sustainable future.

Transition team members are integrated into the local government structure and work alongside influential and willing local stakeholders, such as universities, innovation labs or non-governmental organisations (see Fig. 1). This multi-actor group is entrusted with orchestrating and facilitating co-creation and co-implementation activities, involving various local government bodies, businesses, educational institutions, and community organizations. The team's role is to foster collaboration, trust and alignment among heterogeneous stakeholders, transcending traditional silos and hierarchical models. By acting as a neutral intermediary and facilitating network governance, the Transition Team encourages innovative solutions and empowers local actors to actively shape the future of their city.

In practice, the Transition Team embodies social innovation, reshaping how cities mobilize resources, expertise and collective action to attain climate neutrality. Employing participatory, transdisciplinary approaches, it prioritizes reflexive learning, enabling the city to evolve amid uncertainty and rapid change. The Transition Team serves as a catalyst for local social innovation, igniting transformative change within a community. Operating as a dynamic hub for action and synergies, it bridges diverse sectors and domains, enhancing their ability to collaboratively develop and implement transformative initiatives.

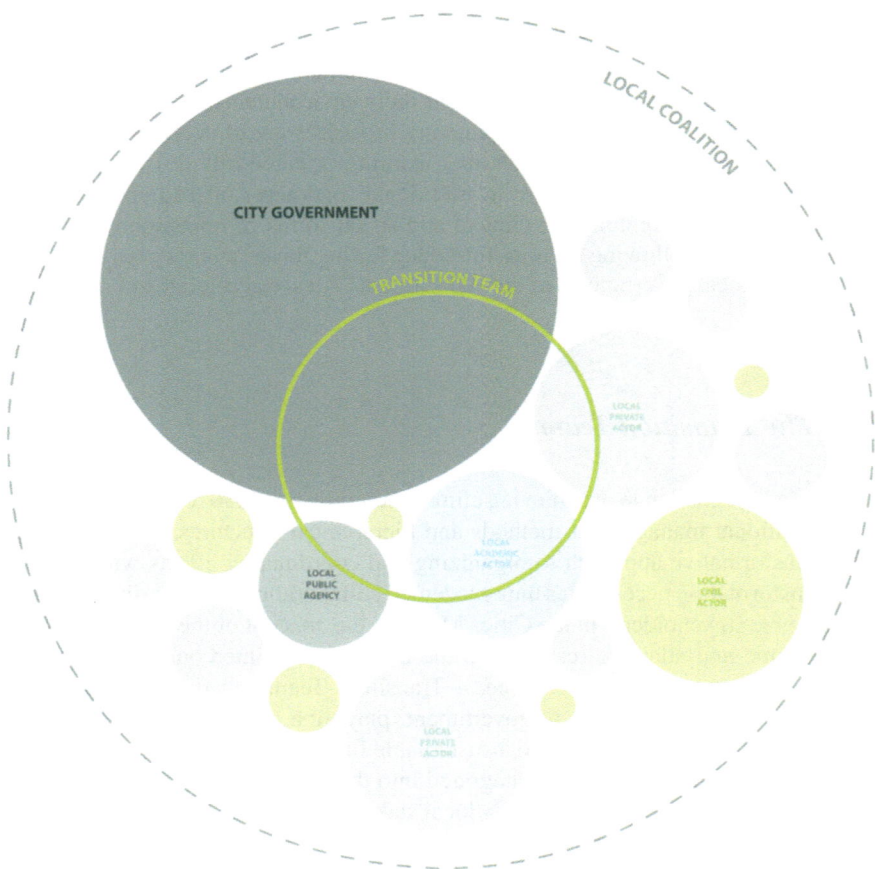

Fig. 1 The transition team in its ecosystem of actors

3.2 Planning Around Portfolios of Actions

Contemporary approaches to systems transformation increasingly revolve around developing portfolios of actions (Chôra Foundation 2021; Danish Design Centre 2023; Gatti and Belle n.d.; Hill 2022; Seppälä 2021; UNDP 2022). These portfolios tackle barriers to change and illuminate impactful pathways forward. They encompass actions spanning social, economic, and technical systems, transcending mere technical innovation to encompass new governance or democratic models, policy changes, financial innovations, diverse business models, and capacity building—essentially embracing various forms of alternative and social innovations. A portfolio's essence lies in how the voluntary integration of these interventions propels acceleration toward a specific objective. For cities, developing a portfolio is advantageous as it prompts a reflective process, aligning numerous scattered interventions and planning strategically toward achieving climate neutrality.

Distinguishing portfolios from traditional roadmaps is crucial. Portfolios of actions are directed toward a vision and expressly work to construct pathways to that vision, primarily by progressing through a series of learning objectives or goals. In simpler terms, a portfolio reflects the characteristics of the problem it addresses. It grapples with the emergent and organic nature of the climate-neutrality challenge through an evolving and organically adapting portfolio, explicitly designed to yield fresh insights into what works and what doesn't as the context responds to interventions. Portfolios embody the essence of social innovation by dynamically testing and de-risking, assigning each action the role of an experiment that validates or uncovers new pathways. Unlike traditional road-mapping or linear action planning, portfolios not only strive for impact but also serve as spaces to explore new possibilities and opportunities.

Portfolio development involves multiple layers of consideration, making it promising for systems change. In the face of multi-layered and dynamic challenges, portfolios reflect this complexity with similarly multi-layered and dynamic approaches, tackling complexity with complexity. To reveal pathways to sustainability, portfolios intentionally overlap multiple layers of consideration: holisticness, synergies, leverage, co-benefits, scalability and replicability. They navigate multidimensional challenges and their multiple levers of change simultaneously, fostering holistic solutions. Portfolios enable synergies between different actions, enhancing financial, data, knowledge, capacity, or asset synergies. Each intervention contributes to elevating other concurrent or subsequent interventions, triggering cascades of changes by addressing structural causes. Effective portfolios target mental models and social change as well as low-hanging fruits, generating direct and indirect positive effects (co-benefits) across environmental, economic, and social indicators. They magnify these effects and cascades of change by emphasizing replicability and scalability, ensuring methods can be applied to various contexts or scaled up effectively.

However, the development and implementation of such dynamic and multi-layered efforts is a novel practice for local governments. In administrative contexts that tend to be divided in sectoral services and departments, subject to multiple overlapping imperatives (service delivery but also imperatives deriving from national and intermediary levels of governance), the development of such portfolios demands an unusual degree of collaboration and mandate, facing an important capacity gap.

3.3 Mobilising a Cross-Sectoral Ecosystem

Developing a democratic approach to achieving climate neutrality, especially through a portfolio strategy, presents a significant challenge when considering that local governments, and even individual administrative services, have direct control over only a fraction of local greenhouse gas (GHG) emissions. Attaining climate neutrality fundamentally involves collaboration. Co-ownership of the effort and shared responsibility for portfolio development and implementation with urban stakeholders are essential. This highlights the need for innovations that facilitate collaborative agenda-setting, the collective formulation of necessary change levers, joint decision-making, and collaborative implementation and evaluation.

Originating in the Swedish context and later adopted by the European Commission for the "100 Climate-Neutral and Smart Cities Mission," the Climate City Contract (CCC) represents a governance model based on a social contract (Shabb and McCormick 2023). According to the European Commission, a CCC entails a firm commitment, an overarching climate action plan detailing a cross-sectoral portfolio, and the related investment plan (European Commission 2021). The CCC serves as both a commitment between local actors, citizens, and national, regional, or European authorities, and a social innovation process in itself. It relies on the engagement and co-creation of urban stakeholders from public, private, civic, media, finance sectors, and citizens to define and implement the action plan.

While the CCC stands as a crucial innovative tool to foster shared ownership, responsibility, and accountability, it challenges the way local governments engage with their territories and urban stakeholders. It shifts the perspective on climate action from a government-owned issue to a territory-owned one. By ensuring social needs are better represented in priorities, it brings about renewed legitimacy and cooperation. However, it also introduces new power dynamics and a bottom-up planning approach that transcends traditional realms of authority.

3.4 Understanding the Way Forward Iteratively

Although historical examples of systems transformation abound (Simms and Newell 2017), the path to climate neutrality remains uncharted territory. Achieving climate neutrality demands a combination of quick wins, utilizing readily available technologies, and experimentation in new domains to refine strategies in the face of uncertain trajectories. Learning through practical, problem-solving interventions is integral to fostering systems innovation (Sengers et al. 2019), allowing for the adjustment of strategies and plans in response to desired outcomes and the surrounding ambiguous environment.

This transition journey necessitates a departure from linear planning. Local governments must adopt new mental models that discard the illusion of top-down road mapping and embrace a humble iterative approach with well-developed experimentation and sensemaking capabilities. It might be challenging for any government, including local ones, to admit that it doesn't possess all the necessary solutions. However, this shouldn't hinder actors from leading the way in collaboratively discovering these solutions over time. The exact nature of challenges and the best ways to address them can be unknowable at the outset—a reality that doesn't necessarily pose a problem as long as there is sufficient agreement, a thin consensus, on the needed direction for learning potential solutions. This approach empowers city governments and multi-actor coalitions to progress with an action-oriented, learning-focused methodology that is both humble and realistic.

3.5 A Tool to Develop and Iterate Multi-actor Action Plans: The Climate Transition Map

In its effort to support 112 European cities of the European Union's *100 Climate-neutral and Smart Cities* Mission, Net Zero Cities has developed a tool to articulate the journey for local governments and the demands it entails: the Climate Transition Map. A key navigation element for cities, the Climate Transition Map (CTM) operationalizes the 3 required shifts for local government to lead transitions: (1) planning around strategic **portfolios** of action (2) in **collaboration** with an inclusive ecosystem of urban stakeholder, (3) in order to **iteratively define** the way forward. It articulates with practicality how these 3 core capabilities shape the day-to-day activities of a **Transition team** in local governments.

The CTM is built around a cycle of 7 phases of work to accelerate the transition: building a strong mandate, understanding of the system, co-creating a portfolio, taking action, learning and reflection, and finally institutionalizing, all while continuously activating an inclusive ecosystem for change (see Fig. 2). In this journey, the CCC is positioned as an instrument of change acceleration, structuring the process of co-creating a portfolio with a multi-actor ecosystem. The journey is cyclical, with local government teams being more than likely to go through the journey multiple times before reaching the climate-neutrality goal. In that, the CTM framework embodies the logic of portfolio, collaboration and iteration.

To work through aach phase in the CTM framework requires the embedding of a number of social innovation capabilities, from experimentation and rapid learning to engaging stakeholders' perspective from the bottom-up and building the ecosystem's creativity and ability to innovate (see Table 1). Together, these capabilities redefine the role of local governments, moving beyond the traditional concept of local service

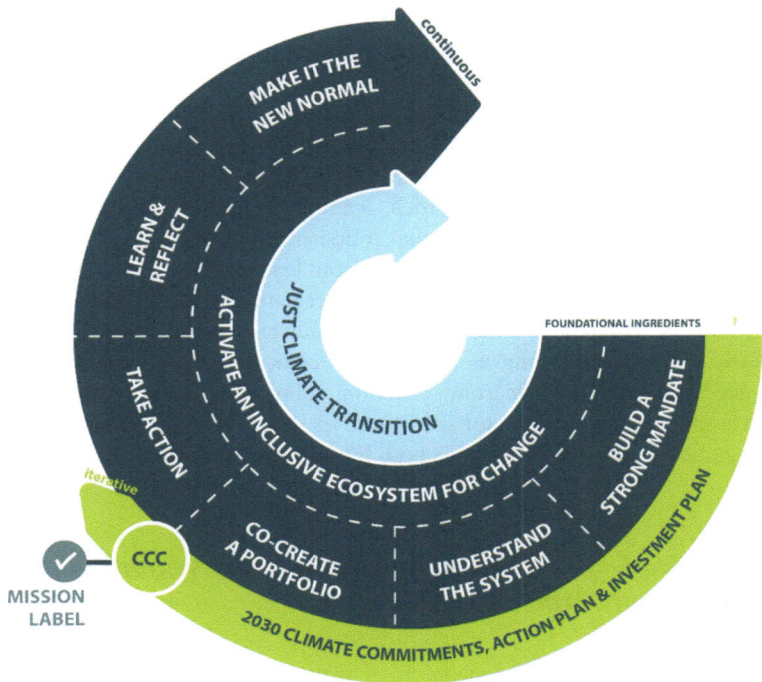

Fig. 2 The NZC Climate Transition Map. *Source* https://netzerocities.app/ClimateTransitionMap

delivery. NZC and the cities it supports have termed this role as "orchestration," drawing parallels to orchestra conductors. Similar to conductors unifying multiple performers, setting the tempo, and modulating the music's pacing, orchestrating a transition involves understanding the vision, implementation dynamics, and effectively communicating these aspects to a diverse ensemble of actors. This practice of orchestration is profoundly social, revolutionizing how local governments engage with their surroundings to tackle challenges.

In summary, the orchestrator vision of local government offers several key advantages driven by social innovation. A Transition Team generates momentum and direction, fostering trust and ensuring communication between actors through robust collaborative practices. Orchestration enables cities to establish inclusive cross-sectoral goals, providing a deeper understanding of root causes and unmet needs. It channels cities' efforts into crafting strategies and projects that harness the knowledge, skills, and perspectives of diverse disciplines and actors from all sectors. This innovative approach to organizational structure, governance, and processes optimizes limited budgets and enhances implementation efficiency by pooling resources and investing strategically at a collective level. Shifting the focus of transition management from a technical concern to a social one, orchestration addresses complex

Table 1 Climate transition phases and the new capabilities required

Activities	Capabilities required
Build a strong mandate	
The local government team needs to be deeply connected across services and departments, with strong relationships at a senior strategic level and at an operational level. Only with such connections can the local government strengthen and build a mandate for action. The full local ecosystem of actors in the city—including citizens, civic groups and the media alongside the private and public sector—needs to be positively engaged in the challenge. Communication, radical collaboration and positive engagement across the political spectrum, including with other levels of government are necessary for the climate agenda to succeed	*Building platforms for collaboration* A strong mandate for change can be the result of the strong legitimacy of the type of change suggested and of those to implement it, of proven impact of the interventions proposed and of confidence that change is possible. In short, interpersonal trust is the basis of shared purpose building (Seitanidi and Crane 2009; Sloan and Oliver 2013) To build momentum, transition teams thus need strong techniques to hear unmet social needs and put urban stakeholders at the centre of action. It is the very process of inclusive collaboration and co-creation which enhances a mandate for transformation
Activate an inclusive ecosystem for change	
The local government plays a key role in developing an inclusive multi-actor ecosystem over time. This requires public servants and engaged participants to create an inclusive setting in which all can contribute their perspectives, be heard, and overcome issues. The development of a shared narrative is crucial to agree on a vision and a common language that creates a shared direction among local actors. This means continuously building trust, proximity and the capability to act of people, particularly those whose experience, imagination and understanding of problems differs from those most commonly involved in decision making	*Enabling inclusive pathways to transformation*: Creating the entry points for diverse stakeholders to take part in climate action demands Transition teams to deploy relational and experiential practices, collaborative methods for creativity and synthesis among them. The ability to govern collectively a shared vision is also essential for a truly distributed social accomplishment *Ecosystem infrastructuring*: To make inclusive collaboration possible takes time and resources, dedicated to continuously creating the space for relations between actors to unfold (Hillgren et al. 2011). Transition teams have a crucial role in nurturing favourable conditions (financial, material, spatial...) for opportunities to emerge in the network
Understand the system	

(continued)

Table 1 (continued)

Activities	Capabilities required
Understanding the challenge of climate-neutrality in all its complexity is not straightforward. Uncovering the barriers that block necessary changes is crucial, facilitating exploration across value chains, sectors and scales, from the micro to the macro, to uncover key interdependencies and entry-points for action. Collective reflection on data and future scenarios, successes and difficulties encountered is to inform the different actors about what choices entail and plausible avenues for transformation	*Responding to emergent needs:* The challenge space of sustainability transitions is ever-evolving. Transition teams will need to bring forth the creative collaboration skills, people-centred mindset and iterative process needed to continuously identify and address the emergent needs of communities as the transition unfolds *Rehearsing the future*: With a problem space also characterised by uncertainty and emergence, techniques of visualisation and materialisation of possibilities (impact, opportunities, unintended consequences) play a central role in supporting decision-making with tangibility, mediating choice
Co-create a portfolio	
The co-creation of a whole city portfolio is an ongoing process that brings together public servants, elected officials and all urban stakeholders. It means combining existing policies, actions and programmes with new interventions into a coherent and strategic set. The co-creation thus demands to align multiple actors' actions into pathways for the transformation of the city and of the life of its people, maximising co-benefits and learning for the collective	*Co-creation*: scoping complementary and reinforcing portfolio actions at a city-scale, rather than municipality-scale, means engaging urban stakeholders at a strategic level of decision-making. To initiate interventions across all priorities and to unlock barriers to change, Transition teams will need to be able to engage stakeholders on a joint responsibility basis *Emergence of value constellations*: portfolios are not only sets of actions but sets of relationships between actors initiating and implementing interconnected interventions. This is a particularly difficult facet of the transition journey: value is created in new forms through the co-design process, implying new forms of relationships and roles that are to be carefully orchestrated (Luján Escalante 2019)
Take action	

(continued)

Table 1 (continued)

Activities	Capabilities required
Practical implementation of a multi-actor portfolio can be confronted with many operational, behavioural and financial uncertainties. Detailed investment planning is a particular focus, which most certainly demands an innovative approach to financing in order to mobilise the scale of capital necessary. Climate action involves a high degree of tactical flexibility in actors' coordination and in experimentation, determining on an ongoing basis interventions' feasibility, impact and social acceptability to improve chances of successful uptake in intervention context	*Change facilitation*: to lead large-scale processes of change with multiple stakeholders, Transition teams will find it important to choreograph collaboration, deliberately planning for the emergence of desired impact (Aguirre et al. 2017). This may rely on design facilitation: the development of context specifics tools and processes to ensure a collective understanding and a shared sensemaking practice *Co-production*: The co-implementation of interventions is central to the delivery of a cross-sectoral portfolio. Collaboration is essential for systemic impact and for people-centred outcomes. This, however, requires for Transition teams to share delivery responsibility with stakeholders
Learn and reflect	
Learning, reflexive practice and adaptive management need to be central and resourced by the local government and the ecosystem of urban stakeholders. Generating data, information and knowledge about actions implies having processes of observation, stock-taking and sharing in place. These processes are the basis for collective sensemaking, reflection and synthesis. Acting upon learning, by pivoting and adaptive decision-making is what makes implementation impactful and resilient	*Experimentation and de-risking*: in a dynamic environment with interdependencies between the technical, social, economic, ecological, arbitrating different possibilities and moving forward relies on action-based learning. Testing hypotheses and making sense of consequences in full context with other actors, experimentation is at the core of understanding and testing solutions' viability, feasibility and desirability
Make it the new normal	
New practices that prove effective are to be embedded, this can include inclusive decision-making, multi-actor collaboration, new budgeting practices or novel mobility business models, techniques for construction. Nurturing networks and communities of practices can support diffusion of change. More effectively, adapting guidance, standards, building capabilities through trainings or mentoring and formalising changes in policy are various avenues for the institutionalisation of a new climate-neutral paradigm	*Scaling across transition levels*: A Transition Team should be in the front-line to instil new, impactful solutions into local culture and individual habits, effectively scaling down these practices to the community level. Expanding the reach of positive practices, amplifying their impact across larger contexts to achieve a greater magnitude of influence—or scaling out—is also a top priority. the Transition Team needs to possess the expertise to institutionalize and codify these innovations through the development of policies, regulations, norms, and standards, thereby scaling up their adoption to a regional or national level (see Moore et al. 2015)

emission challenges, realizing co-benefits, and enhancing coherence between actors, empowering them to take action.

Recognizing the novelty of this perspective for local governments, NZC has developed a Capability Building Programme (CBP) designed to support the development of these orchestration capabilities and the corresponding changes in organizational structure, governance models, and culture. The following section outlines the program developed by NZC to effectively assist cities in embedding these changes in local government, institutionalizing social innovation for climate-neutrality transitions.

4 Learning for Embedding Sustainability and Social Innovation in the City

Tasked with supporting 112 Mission Cities on their journey to climate-neutrality, NZC is dedicated to providing essential resources to enhance local government capacities in harnessing social innovation for climate-neutrality. This support is facilitated through comprehensive learning experiences that offer knowledge, services, and peer-to-peer interactions crucial for embedding social innovation in local government practices. To achieve this goal, NZC has meticulously crafted a Capability Building Programme (CBP), a structured learning journey comprising various modules that guide cities through leading climate action on the ground. The primary objective of the CBP is to facilitate continuous learning and reflection, essential for cities to sustain newly acquired practices and skills in their journey towards climate transitions.

At the heart of any effective capability building initiative lies the formulation of principles and specifications for enduring outcomes. In line with this approach, NZC has innovatively defined the audience and established three core service specifications: action-based learning, community-based learning, and a multifaceted learning journey with multiple touchpoints. These specifications are designed to ensure a lasting and transformative impact.

4.1 Learning for Sustainability

Resistance to change, fatigue, and learning obstacles pose significant challenges when introducing new skills and practices to an organization or ecosystem. Several principles are crucial for facilitating learning and embedding new practices, skills, and structures for the long term. One key strategy is to identify pivotal roles within the local ecosystem. These roles not only set an example and champion the transformation but also diffuse new practices through their actions (Kotter 1995). Leveraging these influential roles can help form coalitions where individuals and teams explore and continuously improve new practices together. Initiating change involves cultivating a culture of openness and adaptability. Encouraging transparent communication of ideas and assumptions, and maintaining open feedback channels, can foster a conducive environment for change (Senge 1997). The following sections delve deeper into each of these principles and explore how they can be integrated into programs aimed at enhancing local governments' ability to lead transformation through social innovation.

An innovative definition of audience. At the heart of NZC's approach to supporting cities in their journey towards climate-neutrality by 2030, the concept of the Transition Team emerges as central not only for facilitating cross-sector collaboration and collective efforts necessary for transformation but also for building capabilities within cities. In its role as an orchestrator, this team occupies a unique position within the local ecosystem, absorbing and disseminating new ideas, practices, and mindsets. Operating as a steward of the local ecosystem, the Transition Team is deeply connected to and influences the local community. This unique position makes the Transition Team an exceptionally effective target audience for a Capability Building Programme. It forms the core of our capabilities diffusion model (refer to Fig. 3), serving not only as a team but also as a network of organizations and, in the most promising cases, as a standalone organization itself.

Action-based learning. Social innovation, akin to many fundamental practices for expediting climate-neutrality, is inherently rooted in practical application. Its value becomes evident through active implementation (Young 2011); its principles—awareness of needs and systems, collaborative governance, co-creation, and experimentation—manifest tangibly through practical application. Emphasizing practical actions and solutions to real challenges is vital for demonstrating the value of new capabilities and generating substantial impact. Therefore, structuring the learning journey around hands-on experience and learning through local practices is essential. Numerous minor changes in everyday activities play a pivotal role in achieving success. These changes can occur at individual, team, and organizational levels. Developing new capabilities, including those in social innovation, demands the adaptability to align closely with a specific local context. This connection serves as a guide for immediate tasks. Learning is rooted in the day-to-day activities and responsibilities of the Transition Team. This approach aids in understanding how incremental changes in capabilities can culminate in a comprehensive citywide pathway to climate neutrality.

Community-based learning. Social innovation presents a unique challenge when introduced into environments that prioritize stability and resist change. Nevertheless, cities across Europe are actively accelerating their progress toward climate neutrality. Many of these cities exhibit remarkable ambition in their climate transition efforts, accumulating a wealth of innovative skills and knowledge in the process. They are keen to share their insights and experiences with neighbouring cities. While city contexts may vary, group dynamics enhance learning from peers: "the likelihood of an individual adopting new norms or institutions increases with the number of neighbours who have adopted them" (Young 2011). Diverse experiences and perspectives encourage the presentation of multiple viewpoints and the challenging of assumptions. Reflective discussions and comparisons are pivotal for stimulating

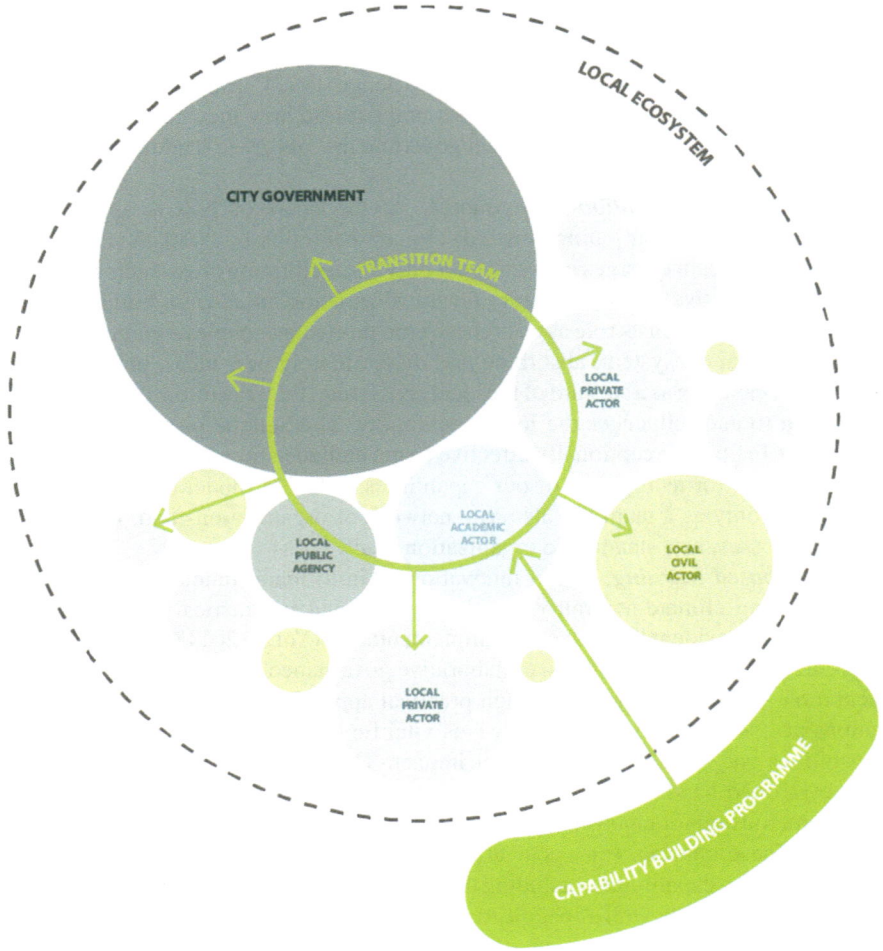

Fig. 3 Capabilities diffusion model

change within individual teams or organizations and beyond. NZC's Capability Building Programme capitalizes on the creation of networking spaces, both online and offline, fostering peer-to-peer connections, exchanges, and collaborations. These platforms serve as arenas for discussions, enabling inspiration, reflection, learning, and adaptation for practitioners in Mission Cities.

Multiple touchpoints learning journey. Social innovation is deeply rooted in prioritizing people and their needs in the process of problem-solving (Mulgan 2006). Developing a social innovation practice involves embracing the diverse needs that coexist within a complex system. This principle also holds true for capability building within the Cities Mission, where cities exhibit a wide range of intrinsic and extrinsic characteristics, coupled with a variety of expressed needs. Mission Cities are at different stages on their journey toward climate-neutrality and encounter a spectrum of barriers influenced by their local, regional, and national contexts. Consequently, there is no one-size-fits-all content package or fixed delivery format that can adequately address the needs of all cities. The Capability Building Programme (CBP) and the Activate the Ecosystem module must accommodate multiple models of learning, various learning styles, and diverse learning goals. This necessitates a range of online and offline learning formats, including expert roundtables, workshops, peer exchanges, and a diverse library of visual, auditory, and kinesthetic learning experiences that cover a wide array of practical questions.

NZC's initiative strives to enhance the capacity of local governments to spearhead transformative change through the adoption and even institutionalisation of social innovation in climate action planning. This entails an approach which breaks down organisational and cultural change into practice-oriented learning blocks directly connected to day-to-day practice and priorities. For this reason, NZC's capability building effort is concentrated around Transition Team's effort to construct their CCC and leverages the Climate Transition Map as a frame of reference, bridging what can be experienced as abstract new practices with everyday preoccupations.

4.2 The Structure of the Modules

Making the most of the value of the Climate Transition Map framework in facilitating work with cities, the NZC Capability Building Programme builds on this framework to organise learning. Intended to support the type of continuous learning and reflection necessary for cities to sustain transformation at scale and speed, the programme suggested isn't meant as a form of linear training but rather as a collection of modules for Transition Teams to lean on and challenge their practice. Each module explores through action a block of necessary capabilities to lead this journey through uncertainty and urgency (see Table 2).

4.3 Continuous Development

Insofar as the current deployment of these modules allows to draw a conclusion, we observe a need to shift local governments' vision of the CCC from a mere compliance exercise to a strategic tool that accelerates the learning process. Local governments often habitually create strategic documents in response to legal requirements, following a rigid multi-level accountability system. Unfortunately, this process frequently overlooks the potential to scaffold learning and transformation within their own organizations and throughout the local ecosystem. Consequently, the deployment of the Climate Transition Map framework by Transition Teams and the utilization of NZC's capability building modules represent just the initial step in institutionalizing social innovation practices for the climate-neutrality transition.

Table 2 Capability building modules for orchestration

CTM phase	Module name	Purpose	Content	Social innovation relevance
Building a strong mandate	Building a shared understanding	Build a shared understanding of the mission and climate city contracts as a multi-stakeholder process	– Building collaborative teams—establishing effective governance structures—stakeholder engagement and buy-in—understanding interactive processes in urban transitions	*Building platforms for collaboration* *Reimagining value constellations* *Facilitating collaborative change*
Activating an inclusive ecosystem for change	Activating an ecosystem for change	Equip participants with knowledge and skills to drive and ignite collaborative and inclusive change locally	– Co-creating a shared vision—engaging communities and stakeholders—maximizing social innovation—futuring techniques and strategic planning—designing spaces for engagement	*Building platforms for collaboration* *Enabling inclusive pathways to transformation* *Ecosystem infrastructuring*
Understanding the system	Setting up a baseline for action and impact	Provide practical support for setting up an impact framework which drives adaptive learning	– Sensemaking—data-driven decision-making—aligning impact logic with indicators and outcomes—effective data visualization	*Responding to emergent needs* *Experimentation and de-risking*
Co-creating a portfolio	Experimenting with a portfolio of action	Equip participants to design coordinated climate action portfolios	Portfolio thinking and strategy—identifying and addressing barriers and gaps—scenario planning and experimentation—cross-departmental and multi-stakeholder co-design	*Co-creation* *Experimentation and de-risking* *Reimagining value constellations* *Scaling across transition levels*

(continued)

Table 2 (continued)

CTM phase	Module name	Purpose	Content	Social innovation relevance
Co-creating a portfolio	Building a strong economic case	Explain the importance of the economic case in understanding carbon impacts and monetary costs/benefits	Economic analysis and costs calculation—prioritizing investments for impact—communicating value to stakeholders—financial modelling	*Rehearsing the future*
Taking action	Developing a financing strategy	Cover financial cornerstones of climate-neutrality investments and financing strategies	– Identifying and planning typical investments—engaging asset owners and financial stakeholders—leveraging financial instruments and solutions—drafting investment plans—crafting financing strategies aligned with city requirements	*Co-production/ implementation*

Our intention is for the Capability Building Programme (CBP) to remain a dynamic and responsive tool for supporting Mission cities. The NZC CBP serves not only as a vital instrument for cities to enhance their capabilities for leading systemic transformations toward climate-neutrality but also as a crucial tool for the NZC consortium and the wider EU Cities Mission ecosystem. It offers valuable insights into the drivers, barriers, risks, and opportunities inherent in climate transitions. From this perspective, we consider it a significant input for its continuous development as well as for the advancement of climate transition initiatives across Europe. It will be pivotal to ensure that conversations with local governments in the various learning modules converge into a continuous learning loop, consolidating the Mission's proactivity in the face of local challenges. Moreover, these discussions open doors for a broader discussion on the imperative of government transformation in European, national, and regional arenas.

5   Empowering Cities for Climate Neutrality: New Capabilities and Their Limits

The journey towards climate neutrality is certainly a complex one that requires substantial changes in the ways local governments use to address public issues. As explored throughout the chapter, this necessity is calling for the exploration of several new notions, including that of transformative policy—such as missions—in alternative to more traditional innovation policies. The central tenet of missions is in fact the possibility to cut across policy domains, institutions and sectors to facilitate system transformation. These new policy types address complex issues, such as the climate transition, that make them open-ended. As such, they also demand a diverse type of governance and coordination that governments at different levels are still exploring (Flanagan et al. 2011). The most thorough study of how nations are carrying out their missions in practice was conducted by the OECD, and it identifies a number of factors and principles crucial to successful implementation, such as strategic orientation (e.g. directionality, intention, legitimacy, flexibility), policy coordination (e.g. horizontality, verticality), and policy implementation (mix of interventions, fundability, evaluability, and learning) (Larrue 2021; Larrue et al. 2019). Because of their novel nature and higher complexity, missions also demand a broader range of capacities and capabilities from public organisations. In the literature, these capabilities are understood as multi-level constructs including individual, institutional and network capabilities (Gieske et al. 2016). At individual and institutional levels, we can include capabilities like sense-making to navigate the system at hand—in our case, all the systems influencing climate neutrality. Orchestration of networked systems of governance is also crucial, while operational capacity is needed for formulation and implementation of activities. These capacities must also be nimble (with flexibility in systems, governance, and finance) to alter direction as needed because of the timeframe involved in missions and the frequently iterative and dynamic character of the mission itself (Weber et al. 1999). This further emphasizes the necessity of constant observation and analysis that incorporates reflexivity and learning into the planning and execution of missions (Molas-Gallart et al. 2021). These requirements clearly request an organization to evolve its capabilities and/or bring in external expertise, a need that NZC recognizes and addresses through the capability building programme and the novel approaches described in the previous paragraphs. In our experience, several are the capacities that local governments are seeking to develop in order to equip themselves better for sustainable transitions. In the following, we discuss the main ones.

Dynamic portfolio management. In the realm of sustainable transitions, development paths are not fixed; they necessitate continuous navigation, sense-making, and the use of suitable frameworks for coordination (Wanzenböck et al. 2020). Navigation in sustainable transitions demands a profound conceptual understanding of the overall system and analytical skills to recognize system conditions, monitor changes, and shape narratives and directions (Edler et al. 2021; Kuhlmann and Rip 2018). This skill resembles effectively using a compass rather than interpreting a static map. Additionally, it involves the ability to curate portfolios of projects and to leverage various funding instruments and policy blends (Kattel and Mazzucato 2018). Furthermore, it relies on agility and mechanisms to alter direction through governance structures and within portfolios (via adjustments in instruments and investments) as necessary (Kivimaa and Kern 2016; Kuhlmann and Rip 2018).

Coordination of multi-actor ecosystems. In the realm of sustainable transitions, it's essential to view the process within a broader organizational context, encompassing various actors contributing to the formulation and execution of these transitions. This demands the leading organization to facilitate collaboration both within and between organizations. Internally, this involves fostering information exchange and building trust by integrating diverse capabilities (cross-disciplinary, sector-specific, specialist, and organizational) to facilitate the diffusion of knowledge and innovation. Achieving this can be facilitated by forming cross-functional teams and transitioning from a traditional hierarchical organizational structure to a more agile and interconnected system, characterized by matrix management and comprehensive accountability. Sustainable transitions also rely on coordination of external collaborations at several scales (local, regional, national). These informal and formal interactions are crucial for establishing the mission's legitimacy and accountability, as well as for achieving wide diffusion and transformative systems change (Wanzenböck et al. 2020).

Learning and reflexivity. Policies and activities related to sustainable transitions are intricate and demand substantial capabilities for learning and adaptability. This involves implementing formal performance management practices, integrating learning and evaluation at both program and transition levels. These practices, in turn, guide the direction of sustainable transitions and the pathways for systemic change. Additionally, fostering creativity within the transition process is crucial. This can be achieved by adopting a double loop learning approach, where continuous learning and reflection are systematically captured, critically engaged with, and then integrated into future practices within the organisation or the broader transition effort.

Overall, in the chapter we have discussed the advantages and the barriers to developing these capacities in local governments to enable climate transitions, while also highlighting the scaffolding and actions that NZC has put in place as a process steward. We have also emphasized the need for transformative policies and innovative approaches which are intrinsic in the concept of missions that cut across policy domains and sectors to facilitate systemic transformation. Finally, the chapter has underscored the evolving nature of sustainable transitions, as requiring continuous learning, adaptability, and creative problem-solving as well as the need to adopt agile, interconnected structures to achieve transformative systems change. NZC's aim is to contribute to shaping the journey to climate neutrality with and by cities and citizens, where a crucial role is played by social innovation in terms of offering a way to go beyond environmental benefits, envisaging new business models, acting within cultural contexts, and empowering collective social action.

References

Aguirre M, Agudelo N, Romm J (2017) Design facilitation as emerging practice: analyzing how designers support multi-stakeholder co-creation. She Ji: J des, Econ, Innov 3(3):198–209

Avelino F, Wittmayer JM (2016) Shifting power relations in sustainability transitions: a multi-actor perspective. J Environ Plann Policy Manage 18(5):628–649. https://doi.org/10.1080/1523908X.2015.1112259

Chôra Foundation (2021) The future of development: "Make Happen" with portfolios of options. https://static1.squarespace.com/static/6358fcee9747a0469c1398a6/t/63934161f956 1c3cca6d08be/1670594918014/Portfolios+of+Options+Green+Paper+16+Mar.pdf

Danish Design Centre (2023) missions lead the way to action in the wilderness of complex problems. https://ddc.dk/missions-lead-the-way-to-action-in-the-wilderness-of-complex-problems/

Edler J, Matt M, Polt W, Weber KM (2021) Transformative mission-oriented STI policy-theoretical and conceptual rationales, intervention logic and challenges of an emerging type of STI policies. In: EU-SPRI 2021: science and innovation-an uneasy relationship? Rethinking the roles and relations of STI policies. https://publications.ait.ac.at/en/publications/transformative-mission-oriented-sti-policy-theoretical-and-concep

European Commission (2021) EU mission: climate-neutral and smart cities implementation plan. https://research-and-innovation.ec.europa.eu/funding/funding-opportunities/funding-progra mmes-and-open-calls/horizon-europe/eu-missions-horizon-europe/climate-neutral-and-smart-cities_en

Flanagan K, Uyarra E, Laranja M (2011) Reconceptualising the 'policy mix' for innovation. Res Policy 40(5):702–713

Gatti L, Belle G (n.d.) Portfolio design Stencils: a conceptual, learning and action architecture for accelerating social systems transformation.

Gieske H, Van Buuren A, Bekkers V (2016) Conceptualizing public innovative capacity: a framework for assessment. Innov J 21(1):1

Grimm R, Fox C, Baines S, Albertson K (2013) Social innovation, an answer to contemporary societal challenges? Locating the concept in theory and practice. Innov: Eur J Soc Sci Res 26(4):436–455

Grin J, Rotmans J, Schot J (2010) Transitions to sustainable development: new directions in the study of long term transformative change. Routledge

Hill D (2022) Designing missions. Mission-oriented innovation in sweden—a practice guide by Vinnova

Hillgren P-A, Seravalli A, Emilson A (2011) Prototyping and infrastructuring in design for social innovation. CoDesign 7(3–4):169–183

Hubert A (2010) Empowering people, driving change: social innovation in the European Union. Bureau of European Policy Advisors (BEPA), 12

Jennings N, Fecht D, De Matteis S (2020) Mapping the co-benefits of climate change action to issues of public concern in the UK: a narrative review. Lancet Planet Health 4(9):e424–e433

Kattel R, Mazzucato M (2018) Mission-oriented innovation policy and dynamic capabilities in the public sector. Ind Corp Chang 27(5):787–801. https://doi.org/10.1093/icc/dty032

Kivimaa P, Kern F (2016) Creative destruction or mere niche support? Innovation policy mixes for sustainability transitions. Res Policy 45(1):205–217

Kotter JP (1995) Leading change: why transformation efforts fail. Harvard Bus Rev

Kuhlmann S, Rip A (2018) Next-generation innovation policy and grand challenges. Sci Public Policy 45(4):448–454

Larrue P (2021) The design and implementation of mission-oriented innovation policies. https://www.oecd-ilibrary.org/content/paper/3f6c76a4-en

Larrue P, Machado D, Yoshimoto T (2019) New mission-oriented policy initiative as systemic policies to address societal challenges: analytical framework and types of initiatives. https://stip.oecd.org/assets/MOIP/CaseStudies/JAP%20SIP.pdf

Loorbach D (2010) Transition management for sustainable development: a prescriptive, complexity-based governance framework. Governance 23(1):161–183. https://doi.org/10.1111/j.1468-0491.2009.01471.x

Loorbach D, Frantzeskaki N, Avelino F (2017) Sustainability transitions research: transforming science and practice for societal change. Annu Rev Environ Resour 42:599–626

Luján Escalante MA (2019) Framework of emergence: from chain of value to value constellation. CoDesign 15(1):59–74

Markard J, Raven R, Truffer B (2012) Sustainability transitions: an emerging field of research and its prospects. Res Policy 41(6):955–967

Molas-Gallart J, Boni A, Giachi S, Schot J (2021) A formative approach to the evaluation of transformative innovation policies. Res Eval 30(4):431–442

Moore M-L, Riddell D, Vocisano D (2015) Scaling out, scaling up, scaling deep: strategies of non-profits in advancing systemic social innovation. J Corp Citizenship 58:67–84

Mulgan G (2006) The process of social innovation. Innovations 1(2):145–162

Nilsson W (2023) The future of teaching and learning social innovation (SSIR). Stanf Soc Innov Rev 21(2):52–54

Piketty T (2022) El capital en el siglo XXI. Fondo de cultura económica

Reynolds S, Gabriel M, Heales C (2016) Social innovation policy in Europe: where next? https://media.nesta.org.uk/documents/social_innovation_policy_in_europe_-_where_next.pdf

Seitanidi MM, Crane A (2009) Implementing CSR through partnerships: understanding the selection, design and institutionalisation of nonprofit-business partnerships. J Bus Ethics 85:413–429

Senge PM (1997) The fifth discipline. Meas Bus Excell 1(3):46–51

Sengers F, Wieczorek AJ, Raven R (2019) Experimenting for sustainability transitions: a systematic literature review. Technol Forecast Soc Chang 145:153–164

Seppälä M (2021) Radical uncertainty requires radical collaboration. Stepping stones towards systems transformation with innovation portfolios. Sitra. https://www.sitra.fi/en/publications/radical-uncertainty-requires-radical-collaboration/

Shabb K, McCormick K (2023) Achieving 100 climate neutral cities in Europe: investigating climate city contracts in Sweden. Npj Climate Action 2(1):6

Simms A, Newell P (2017) How did we do that? The possibility of rapid transition. https://steps-centre.org/wp-content/uploads/2017/04/How_Did_We_Do_That.pdf

Sloan P, Oliver D (2013) Building trust in multi-stakeholder partnerships: critical emotional incidents and practices of engagement. Organ Stud 34(12):1835–1868

UNDP (2022) System change: a guidebook for adopting portfolio approaches. https://www.undp.org/sites/g/files/zskgke326/files/2022-03/UNDP-RBAP-System-Change-A-Guidebook-for-Adopting-Portfolio-Approaches-2022.pdf

UNECE (2016) The co-benefits of climate change mitigation. https://unece.org/DAM/Sustainable_Development_No._2__Final__Draft_OK_2.pdf

Wanzenböck I, Wesseling JH, Frenken K, Hekkert MP, Weber KM (2020) A framework for mission-oriented innovation policy: alternative pathways through the problem–solution space. Sci Public Policy 47(4):474–489

Weber M, Hoogma R, Lane B, Schot J (1999) Experimenting with sustainable transport innovations. A workbook for strategic niche management. https://www.applypedia.com/dep/construction/urbanism/assets/files/Experimenting-withSustainableTransportInnovations._compressed.pdf

Young HP (2011) The dynamics of social innovation. Proc Natl Acad Sci 108(supplement_4):21285–21291

Social Innovation Scaling at Urban Level

Sabrina Bresciani and **Cyril Tjahja**

Abstract How can social innovation initiatives in cities be scaled to provide systemic and enduring impact on reducing GHG emissions? Academic and practitioners' literature offers a number of conceptualizations of social innovation scaling strategies, including the popular "scaling up, out and deep" framework. In this chapter, scaling methods are presented to support cities to amplify social innovation impact. Based on cities' social innovation readiness level, social innovation pathways can range from preparatory phases—including capacity building and mobilizing funds—to mapping the ecosystem, and to systematically embedding social innovation for climate neutrality in all cities' activities, including communication strategies, public procurement, circular economy, urban planning, and policy co-creation.

Keywords Scaling strategies · Pathways · Systemic innovation

Stimulating social innovation on a city level is regarded to be particularly beneficial for reducing greenhouse gas emissions (Coutard and Rutherford 2010). Local governments can act as intermediary actors, helping energy communities and other grassroots social innovation initiatives to initiate new projects and local networks, provide tools, resources and professional services, manage and evaluate funding programs, and connect these initiatives to policymakers and other relevant stakeholders (Hargreaves et al. 2013).

However, expanding the impact, or scaling up the activities of promising local initiatives, is a challenging task. For the initiatives, it is difficult grow independently when not encapsulated in an *enabling ecosystem*, which can provide them with social, cultural and technological support (Manzini 2015). Although cities could

S. Bresciani (✉)
Department of Design, Politecnico di Milano, Milan, Italy
e-mail: sabrina.bresciani@polimi.it

C. Tjahja
TNO and Vrije Universiteit Amsterdam, Amsterdam, The Netherlands
e-mail: cyril@schonevormen.nl

© The Author(s) 2025
S. Bresciani (ed.), *Social Innovation Projects for Climate Neutral Cities*,
PoliMI SpringerBriefs, https://doi.org/10.1007/978-3-031-87726-1_4

play a pivotal role in facilitating or stimulating such ecosystems, public administrators might not have the manpower and/or expertise available in-house to provide this support. Several scaling strategies have been identified in the NetZeroCities project that could be beneficial for cities, to help them make informed decisions that can provide meaningful support to local social innovation initiatives, and to embed social innovation in a systemic way into a city's plans.

1 Strategies for Scaling Social Innovation

In the past decade, a number of approaches have been developed by scholars and practitioners, which could be useful for cities to consider when scaling up social innovation initiatives. In this chapter, we provide an overview of four popular frameworks: (1) *Scaling up, out and deep* (Moore et al. 2015), (2) Nesta's *Make it big framework with four routes to scaling social innovations* (Gabriel 2014), (3) The World Health Organization's *Scaling up framework* (2010), and (4) *Mapping the ecosystem with a city observatory* (Andion et al. 2022).

A review of each scaling strategy and related illustrative cases is provided next.

1.1 Scaling Up, Out and Deep

Moore et al. (2015) distinguish three types of strategies for scaling impact: scaling up, out and deep (https://netzerocities.app/resource-3953).

When *scaling out*, initiatives or organizations use replication and diffusion to expand to a larger geographic area through two main strategies. First, by deliberate replication, which entails spreading or replicating programs both numerically and geographically, while maintaining the core principles of the innovation. Second, by spreading the innovation's principles, but adapting it to local contexts through co-creating knowledge, social media and educational platforms.

Initiatives that are *scaling up* operate on the assumption that social issues go beyond their local contexts, and should be addressed by modifying their systemic or institutional basis. Efforts should therefore be focused on advocating for change or attempting to change or develop new legal frameworks or policies. To scale up, partnerships for advocacy can be developed to strengthen the effectiveness.

Scaling deep entails changing people's cultural norms, beliefs and relationships by generating and spreading 'big' cultural ideas, thereby consciously reframing dominant narratives on particular issues. For example, through sharing knowledge and practices through learning communities, learning platforms and participatory approaches. In addition, investing in transformative learning can be used to promote the innovation to other organizations, engraining its associated beliefs and practices. Methods that can be used to facilitate this are mentorship, sharing organizational or community culture, and shared reflection and evaluation, among others.

Moore et al. indicate initiatives tend to focus on scaling out in their beginning phase, whereas when more significant system change needs to take place, initiatives shift more towards scaling up and/or scaling deep.

1.2 Nesta's 'Make It Big' Framework: Four Routes to Scaling Social Innovations

The UK-based social innovation agency Nesta has developed a framework which outlines four routes that can be followed to construct a scaling strategy (Gabriel 2014): *influence and advise, build a delivery network, form strategic partnerships,* and *grow an organisation to deliver* (https://netzerocities.app/resource-3852).

The *influence and advise* route is recommended for social innovations that are based on principles or utilize methods that can be deployed in different contexts and/ or social innovations that challenge the prevailing norms. As there might not be direct contact between the innovator and those that are using the innovation, the amount of control of how it is implemented is low. However, the advantage of this route is that it can potentially reach a wide audience.

The *build a delivery network* route is suitable for social innovations which feature elements that consist of similar activities, such as transferring knowledge or community building. The innovations or practices are spread by 'delivering' them to their audiences through a network which focuses on replication or in some cases social movements. The focus here lies on a creating a common feeling of purpose, while at the same time ensuring that the innovation stays truthful to the original idea. Examples of these networks are federations and communities of practice.

Scaling through partnering with other organizations is the main characteristic of the *form strategic partnerships* route. In doing so, certain new technologies and skills, which would take a long time to develop, become instantly accessible. Being taken over by government organizations or partnering with large multinationals are common examples of this route. Although this type of scaling provides an opportunity for social innovation initiatives to scale up quickly, it can require some adjusting in order to become successful.

The *grow an organisation to deliver* route is the most suitable choice for organizations which rely on (individual) knowledge that cannot be replicated easily. This route entails growing the organization itself and therefore direct control over the spread of innovation is possible. For larger organizations, this would entail scaling through existing channels and sources, whereas smaller organizations would need to diversify their activities, which might be more challenging from a management point of view.

The four routes do not necessarily have to be followed individually, at times they can overlap or some routes can be followed simultaneously, with some social innovators choosing to scale their organization using all four routes at the same time. The core insights of Nesta's framework is the acknowledgement that scaling is not just about growing an organization, but rather growing its impact. The framework highlights that scaling up requires social innovators to work in a different way, requiring competences in change management.

1.3   WHO Scaling Up Framework

Drawing from related fields, the scaling up framework developed by the World Health Organization (WHO) aims to facilitate the planning and management of the upscaling process and is based on literature and on WHO's extensive international experience (World Health Organization WHO 2009, 2010; https://netzerocities.app/resource-998)).

The framework is an actionable tool for cities to consider main issues central to scaling up social innovations and other types of people-based innovations for climate neutrality. It views scaling up as a system of interrelated elements and strategic decisions that have to be made: the (social) innovation to be scaled up, the institution/organizations that will adopt and implement it on a larger scale, the external conditions and institutions that will affect the prospects for scaling up (environment), and the individuals and organizations that will promote and facilitate wider use of the innovation (resource team). Once these elements have been identified, the strategy can be designed, which entails determining the resources necessary in terms of finances and capacities, the aspects of the process, such as the scope, speed, and degree of participation as well as the method of dissemination, monitoring and evaluation.

More specially, the WHO scaling framework consists of nine steps (WHO 2009): (1) Planning actions to increase scalability and innovation; (2) Increasing the capacity of the user organization to implement scaling up; (3) Assessing the environment and planning actions to increase the potential for scaling up success; (4) Increasing the capacity of the resource team to support scaling up; (5) Making strategic choices to support vertical scaling up (institutionalization); (6) Making strategic choices to support horizontal scaling up (expansion/replication); (7) Determining the role of diversification; (8) Planning actions to address spontaneous scaling up; (9) Finalizing the scaling strategy and identifying next steps (WHO 2009).

The WHO framework has been used to plan and assess health service innovations, and can provide a systematic and holistic view on the elements that can be considered, helping to identify the actions and decision that can be made to designing a strategy for scaling up. It was developed in conjunction with NGOs and international agencies (2010) for the purpose of providing a tool that can provide a structured, systemic view of scaling up.

1.4   Social Innovation Observatory

Before cities can determine which initiatives to scale up or how to support them in doing so, it is beneficial to have a comprehensive overview of the current social innovation initiatives within a city. A *social innovation observatory* can help cities to map and analyze local social innovation ecosystems (https://netzerocities.app/res ource-3842). In Florianópolis, Brazil, such an observatory has been constructed in the form of a collaborative digital platform (Andion et al. 2022). Figure 1 provides a screenshot of the digital map created by the social innovation observatory of the city (https://observafloripa.com.br/reports#toppage).

Social innovations are geographically mapped using a variety of sources, including public programs for the promotion of social innovation, and by inviting the city's social innovation actors to map themselves and their network. With such mapping, network and organizational data can be analyzed for informing policies and measure initiatives' results.

The observatory development process in Florianópolis started by investigating the current laws, regulations, policies and public programs regarding social innova-tion. This was supplemented by interviews with the main actors within the city that support social innovation as well as collecting publicly accessible information about social innovations. In addition, site visits to the initiatives were conducted in order to gain insight in the issues they face, their proposed solutions, their methods and technologies used, and how they measured their results, among others.

Not only innovations were mapped but also support organizations, which belonged to ten categories: (1) training centers; (2) promotion of social entrepreneurship; (3) research and extension centers; (4) funders; (5) technical supporters; (6) articulators

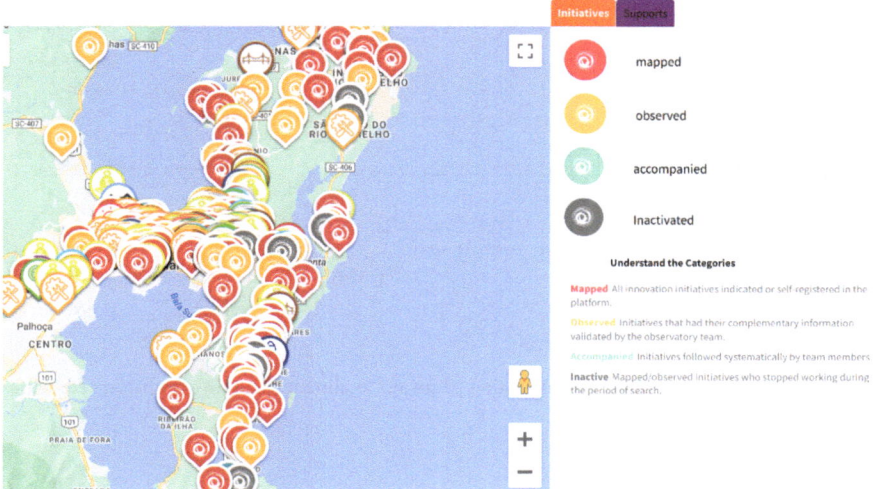

Fig. 1 A map from the Florianopolis' social innovation observatory

and bridging actors; (7) spaces for communication, dialogue, and reflection; (8) incubators; (9) accelerators; and (10) certifiers.

Cities can develop a geographical map of social innovation initiatives with available technologies, such as Google Maps API; the methodology deployed for developing the Florianopolis social innovation observatory is further described by Alperstedt and Graeff (2019).

2 Pathways for Scaling

The above reviewed frameworks can provide cities with a comprehensive and informative overview of scaling strategies. Which scaling routes are the most appropriate depends on a city's readiness level, available resources and long-term objectives. As presented in Chap. 2, cities can act according to three main profiles: the enabler, the mediator and the visionary. To equip cities with a compact but comprehensive map of social innovation routes to scale, we developed the NetZeroCities *Social Innovation Actionable Pathways*, composed of ten subsequent categories according to three phases: prepare, act and scale. Figure 2 shows an interactive map of the categories along the phases, and specifies related case studies and indicators available on the NZC platform (https://netzerocities.app/resource-4074).

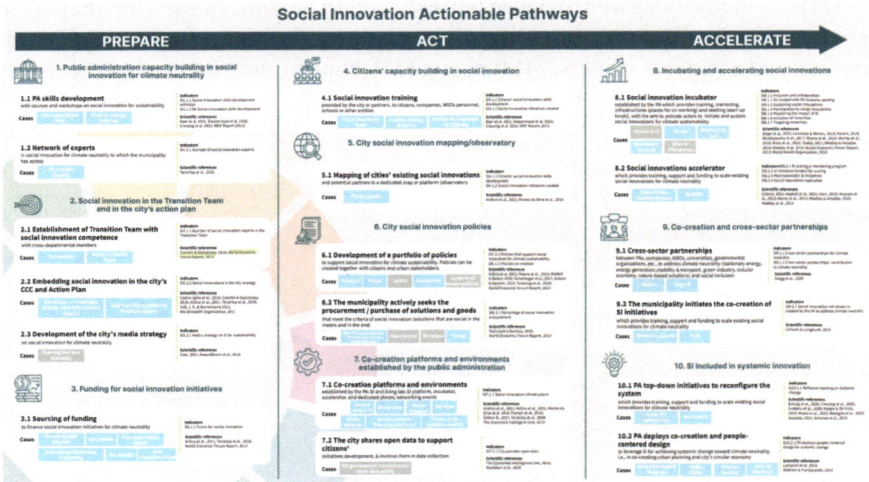

Fig. 2 Interactive graph of Social Innovation pathways to climate neutrality (https://netzerocities.app/resource-4074)

2.1   Phase: Prepare

Category 1: Public administration capacity building in social innovation for climate neutrality is composed of two subcategories: (1.1) PA skills development with courses and workshops on social innovation for sustainability and (1.2) Network of experts in social innovation for climate neutrality to which the municipality has access. Case studies that illustrate these respective categories include City experiment funds, Positive Energy Districts (presented in Chap. 5), and City-studio (Spain).

Category 2: Social innovation in the Transition Team and in the city's action plan is composed of (2.1) Establishment of Transition Team with social innovation competence with cross-departmental members, (2.2) Embedding social innovation in the city's CCC and Action Plan and (2.3) Development of the city's media strategy on social innovation for climate neutrality. Examples of these principles are outlined in the case studies PentaHelix (presented in Chap. 5), Barcelona co-creating a climate plan with citizens (Spain), the Just Transition Listening Platform (Spain) and Framing the Sun (Rosenbloom et al. 2016).

Category 3: Funding for social innovation initiatives, is composed by the subcategory (3.1) Sourcing of funding to finance social innovation initiatives for climate neutrality, illustrated by the cases Sonnet Bristol City lab; You decide; Antwerp participatory budgeting; Mannheim; Just Transition Fund; City experiment funds.

2.2   Phase: Act

Category 4: Citizens' capacity building in social innovation is composed by the subcategory (4.1) Social innovation training provided by the city or partners, to citizens, companies, NGOs personnel, schools or other entities. Cases that illustrate this concept are City experiment funds and Positive Energy Districts.

Category 5: City social innovation mapping/observatory included the subcategory (5.1) Mapping of cities' existing social innovations and potential partners in a dedicated map or platform (observatory). The case study of Florianopolis observatory illustrated in Sect. 4.1.4, is a relevant example for this category.

Category 6: City social innovation policies is composed of the two sub-categories (6.1) Development of a portfolio of policies to support social innovation for climate sustainability. Policies can be created together with citizens and urban stakeholders and (6.2). The municipality actively seeks the procurement/purchase of solutions and goods that meet the criteria of social innovation (solutions that are social in the means and in the end). Cases that illustrate these concepts are: Bologna, PentaHelix, Spain food waste policies, and Oslo public procurement for innovative nature-based solutions.

Category 7: *Co-creation platforms and environments established by the public administration* has two sub-categories: (7.1) Co-creation platforms and environments established by the PA: SI and living lab; SI platform, incubator, accelerator, and dedicated places; networking events, and (7.2) Open data sharing. Cases that illustrate these categories are: Bristol City Lab; Mannheim; Nappi Naapuri; Bologna; Just transition listening platform and the UK government transparency open data policy.

2.3 Phase: Accelerate

Category 8: Incubating and accelerating social innovations is composed of (8.1) Social innovation incubator established by the PA which provides training, mentoring, infrastructures (places for co-working) and seeding (start-up funds), with the aim to activate actors to initiate and sustain social innovations for climate sustainability and (8.2) Social innovations accelerator which provides training, support and funding to scale existing social innovations for climate neutrality. Relevant case studies for this category are: Torino, Bristol City Lab, Mannheim City Lab, Clean Cities ClimAccelerator and the Wiener Klimateam, which is illustrated in Chap. 6.

Category 9: Co-creation and cross-sector partnerships comprises the subcategory (9.1) cross-sector partnerships between PAs, companies, NGOs, universities, governmental organizations, etc., to address climate neutrality (stationary energy, energy generation, mobility & transport, green industry, circular economy, nature-based solutions) and social inclusion, and (9.2) the municipality initiates the co-creation of SI initiatives for climate neutrality together with citizens, local companies, NGOs or other local organizations, to address climate neutrality (stationary energy, energy generation, mobility & transport, green industry, circular economy, nature-based solutions) and social inclusion. Relevant illustrative cases include Malmo; Zagreb, Better Reykjavik and KLIK (reviewed in Chap. 5 of this book).

Category 10: SI included in systemic innovation is composed of (10.1) public administration top-down initiatives to reconfigure the system to support climate neutrality through a portfolio of social innovation initiatives (i.e., urban spaces design, circular economy, etc.), and (10.2 PA deploys co-creation and people-centered design to leverage SI for achieving systemic change toward climate neutrality, i.e., in co-creating urban planning and city's circular economy). Cases that illustrate these principles include Paris 15 min city, Mannheim, Viable Cities, Blok 19 Renewal Program, Vitoria-Gasteiz, and Wiener Klimateam—discussed in Chap. 6 of this book.

3 Reflexivity

Are social innovation interventions always successful? A reflexive attitude is crucial for cities—as well as for theory development—to be able to learn from the in-progress implementation of projects. Qualitatively and quantitatively assessing of social (and other) innovations' contribution to climate neutrality has the benefit of providing internal decision makers as well as the general public with evidence of the progresses or issues to be adjusted. The NZC comprehensive indicators set (Neumann et al. 2023) can not only provide cities with a list of GHG and non-GHG indicators, but also with a list of qualitative reflexive, guiding questions (in the Appendix of the publication as well as in Mureddu and Bresciani 2022), which cities can select to support their internal sensemaking. Ten of the referenced case studies are reviewed in Chap. 5 of this book, while the remaining cases are available on the NetZeroCities portal (https://netzerocities.app/resource-4074). An extended explanation of all categories and related scientific reference and indicators of the social innovation actionable pathways is provided in Bresciani et al. (2023).

References

Alperstedt GD, Graeff J (2019) Social innovation ecosystems and cities: co-construction of a collaborative platform. In: Hoowaldt J, Kaletka C, Schroder A, Zirngiebl M (Org.). Atlas of social innovation, 2aed., vol 1 TU Dormunt University, Dormunt, pp 125–128

Andion C, Alperstedt GD, Graeff JF, Ronconi L (2022) Social innovation ecosystems and sustainability in cities: a study in Florianópolis, Brazil. Environ Dev Sustain 24:1259–1281

Bresciani S, Tjahja C, Komatsu T, Rizzo F (2023) Social innovation for climate neutrality in cities: actionable pathways for policymakers

Coutard O, Rutherford J (2010) Energy transition and city–region planning: understanding the spatial politics of systemic change. Technol Anal Strateg Manag 22(6):711–727

Gabriel M (2014) Making it big: strategies for scaling social innovation. Nesta, London

Hargreaves T, Hielscher S, Seyfang G, Smith A (2013) Grassroots innovations in community energy: the role of intermediaries in niche development. Glob Environ Chang 23:868–880

Manzini E (2015) Design, when everybody designs: an introduction to design for social innovation. MIT Press

Moore M-L, Riddell D, Vocisano D (2015) Scaling out, scaling up, scaling deep strategies of nonprofits in advancing systemic social innovation. J Corp Citizenship 58:67–84. https://doi.org/10.9774/GLEAF.4700.2015.ju.00009

Mureddu F, Bresciani S (2022) Report on Indicators & assessment methods for social innovation action plans. Deliverable 2.7. NetZeroCities EU H2020 Grant Agreement n°101036519. European Union, Brussels

Neumann et al 2023. Neuman et al. (2022) Concept for a comprehensive indicators framework. NetZeroCities Deliverable 2.4.1

Rosenbloom D, Berton H, Meadowcroft J (2016) Framing the sun: a discursive approach to understanding multi-dimensional interactions within socio-technical transitions through the case of solar electricity in Ontario, Canada. Res Policy 45(6):1275–1290

World Health Organization (2009) Practical guidance for scaling up health service innovations. World Health Organization. https://expandnet.net/PDFs/WHO_ExpandNet_Practical_Guide_published.pdf

World Health Organization (2010) Nine steps for developing a scaling-up strategy. World Health Organization. https://apps.who.int/iris/bitstream/handle/10665/44432/9789241500319_eng.pdf

Ten Social Innovation Case Studies to Address Cities' Challenges: Citizens Engagement, Energy and Behavioral Change

Sabrina Bresciani[ID]**, Rohit Mondal**[ID]**, Kaisa Schmidt-Thome**[ID]**, and Francesco Michele Noera**[ID]

Abstract How are cities implementing social innovation for climate neutrality in practice? In this chapter, 10 case studies are summarized according to three main categories: citizen engagement, energy (communities), and behavioral change. These categories correspond to the NZC curated collections of social innovation case studies, which are available on the NZC platform. The breadth of cities' experiences covered by the cases reflect Europe's dynamic experimental environment, with diverse levels of readiness and understanding of social innovation for sustainability. Cases outline cities' experiences from Spain, Greece, Croatia, Belgium, Finland, Netherlands, Belgium, Austria, U.K. and U.S. Topics range from gamified approaches to co-creation and from community building and to civic design labs.

Keywords Case studies · Social innovation · Sustainability · Europe · Energy communities · Citizens engagement · Green nudges · Design for transitions · Pro-environmental behavior

1 How Cities Are Addressing the Climate Neutrality Challenge

Within the NetZeroCities EU-funded project, 40 case studies illustrating the power of social innovation for climate neutrality are developed and provided on the NZC online portal. Cities can learn from the experiences outlined in the case studies,

S. Bresciani · R. Mondal (✉)
Department of Design, Politecnico di Milano, Milan, Italy
e-mail: rohit.mondal@polimi.it

K. Schmidt-Thome
Demos Helsinki, Helsinki, Finland

F. M. Noera
Aalborg University, Aalborg, Denmark

learn how to replicate them and extract learnings. The case studies cover a variety of topics and social innovation readiness levels, from facilitation to systemic change. This chapter provides an overview of a sample of cases categorized into three relevant topics: citizens' engagement, energy and behavioral change. All cases are discussed from the social innovation perspective, and are available in more extended format in the NetZeroCities portal as "curated collections" of case studies (https://netzeroci ties.app/resource-4074). From a theoretical perspective, such a rich source of cases provides evidence that social innovation is already a successfully utilized lever of positive change in cities, within and beyond Europe; they breadth of the cases outline the scalability of social innovation, as discussed in Chap. 2, which can not only enable grassroot initiatives or networking, but are able to support systemic changes through concerted effort which is social and sustainable in the means and in the ends.

2 Citizens' Engagement

In the realm of social innovation, there is a special, cozy place for citizen engagement—and vice versa. While social innovation can introduce new ways of thinking and collaborating to address the most pressing challenges cities face, citizen engagement can support finding those new ways as well as validating them (Anttiroiko 2016; Terstriep et al. 2020; Eckhardt et al. 2021). This Section focuses on the aspects that city organisations might like to explore in the context of their climate action: how social innovations add to the quality and meaningfulness of citizen engagement.

As discussed in Chap. 2 of this book, a city can take various roles with regard to its innovation landscape. Whether a city operates as an *enabler*, *mediator*, or as a *visionary*, would affect what it hopes to achieve through citizen engagement. The Enablers would encourage individual initiative and self-driven participation, Mediators would facilitate structured channels for citizen input, and Committed change-makers would lead transformative efforts where citizens were warmly welcome or even a prerequisite of success. The social innovation projects chosen for this Section _SynAthina (Greece), Civic Design Lab of Oakland (U.S.) and the Just Transition Listening Platform (Spain) - follow this "ladder".

SynAthina (https://netzerocities.app/resource-2856) is a social innovation platform, run by the City of Athens, to support execution of citizen ideas for better city life. Citizens and community groups can submit innovative ideas and get then connected to relevant actors (government representatives, NGOs and private businesses) that support their efforts. The support can include funding, but another incentive for participation is access to the larger network of innovative groups.

According to The GovLab (2023), Synathina was initiated in 2013 and started first as a physical map of grassroots communities in the city, then as a simple online map. With a major grant from the Bloomberg Mayors Challenge in 2014 it was developed into a fully-fledged online platform, and later it has been a part of an Urban Innovative Action funded by the European Commission. SynAthina has been managed by a five-member team within the Vice Mayor's Office, including two designated Public Engagement Officers. The team forms and maintains relationships with community groups through events and other face-to-face gatherings. Besides gathering and implementing solutions in a broad variety of themes, synAthina has also helped to build trust between citizens and the city's government: people have not only been heard but have seen their suggestions taken seriously and their contributions leading to something tangible.

In Synahtina, the city has acted mainly as an Enabler as it has learnt to cherish the creativity and proactivity of its citizens and to crowd-source promising proposals. Although it started by inviting individual suggestions rather than cultivating the ecosystem more broadly, the durability of the platform has supported interconnections between projects and actors, pointing to the Mediator role of the city/platform. Furthermore, as SynAthina welcomes any kind of ideas for a closer scrutiny, so its commitment to more specific missions is not as explicit as with Committed change-makers. In any case, Synathina is a great example of social innovation in action. By being curious and receptive to insights the City of Athens can enjoy the collective intelligence of its inhabitants who, in turn, see the benefits in improved city life.

The **Civic Design Lab** (https://netzerocities.app/resource-2387) of Oakland, in the U.S., facilitates dialogues and co-creates solutions in collaboration with government agencies, local businesses and communities for improved public services. It helps local government decision making by designing current and future policies with the people, for the people and by the people. CDL can be characterised as people-centric innovation lab, where systems thinking and racial equity lens meet human centered design.

The Civic Design Lab (CDL) was established as a result of a public–private partnership to address civic resiliency challenges in Oakland, California. CDL's initiatives and projects are designed for sustained, thematic discussions over the long term. This approach involves creating working groups for each project, ensuring ongoing engagement of citizens and stakeholders throughout all phases, including feedback and evaluation. The addressed topics include rent adjustments, healthy housing and financial empowerment.

The key principles include discovery and discussion. Discovery is about identifying and listening to key voices from local communities, innovators and stakeholders that can influence decision making. Discussion in turn is about bringing together all stakeholders impacted by the problem to continuous, cyclical discussions in varying sizes of groups to facilitate efficient interactions. The discussions are documented, analysed and the data is validated together. The ideas and prototypes that emerge are shared with all stakeholders involved in the process.

In the Civic Design Lab, the City of Oakland acts mainly as a Mediator, connecting the dots between different stakeholders that can address the local challenges together. It does not focus on a certain mission but accepts the priorities that emerge from the discussions. Its approach is based on collaboration that takes the different starting points in terms of capabilities and resources into account, which is a key feature of a Committed change-maker. The **Just Transition Listening Platform** (https://net zerocities.app/resource-3106) has fostered territorial transformation in the mining region of northern Spain (Lada and Velilla towns) since 2020. It has introduced new forms of diagnosis, co-creation and sense-making to see how people are adjusting to the changes in their life and connected their future aspirations with a portfolio of initiatives. In Lada, new opportunities are sought in areas such as energy, circular economy, or green jobs, whereas in Velilla, the attention has focused on food and its role in culture and tourism as well as on industry linked to energy.

Actors behind the platform include the energy company Iberdrola, the Centre for Innovation in Technology for Human Development of the Polytechnic University of Madrid (itdUPM) and the Agirre Lehendakaria Center (ALC). It has fostered collaboration between citizens, public entities and companies to support a just transition that is required following the closures of Iberdrola's coal-fired power plants.

The methodology is based on community listening and co-creation of interconnected projects and initiatives. The platform has also articulated an image of the future where the sustainability transition has not aggravated but diminished the social cleavages of the region by building on the available and further strengthened capabilities and resources.

In the case of Lada and Velilla, the regional cooperation has many features of a Committed change-maker: there is a clear sense of joint direction coupled with a commitment to the fairness of the transition. The cooperation has also some features of a Mediator, as the fostering of interconnections has made it possible for citizens and representatives from companies and other institutions to get to know each other and ideate collaboration possibilities.

These three examples share the two features that are commonly associated with social innovation: "the satisfaction of human needs and the transformation of social relations" (Parés et al. 2017, 5). Some of them touch also upon power relations and transforming them from below, as a third key feature of social innovation (Galego et al. 2022). While none of the presented examples have been put up with the explicit aim to foster climate change mitigation at the local level, they all showcase how social innovations can add to the quality and meaningfulness of citizen engagement, also in the context of cities' climate action.

3 Energy

The following section outlines four strategies to bring forth energy communities and energy-saving behaviors. The first two cases highlight the importance of simplifying the engagement process of citizens through a single touchpoint for citizens. EcoHouse Antwerp in Belgium shows the efficacy of establishing a one-stop-shop that unburdens citizens' decision-making about starting building renovations. The Energy cooperative KLIK (Križevci Climate Innovation Laboratory), in Croatia, brings forth the same strategy, demonstrating the possibility of direct management of change by citizens themselves. The Valencia City Council (Spain) highlights an organizational innovation strategy that incentivizes establishing a network of energy communities on the city territory. Eventually, the SONNET City Lab in Bristol (U.K.)

suggests an additional engagement strategy based on co-designing crowdfunding investment instruments for citizens.

EcoHouse   Antwerp   (https://www.antwerpen.be/nl/overzicht/ecohuis-antwer pen/nieuws; https://netzerocities.app/resource-2813) is a project active since 2015 and part of Antwerp's Climate Plan 2030. It consists of a one-stop-shop that offers services for adopting energy-efficient behaviors and realizing building renovations. EcoHouse services include advice, free installation of energy-saving products, personalized audit and investment plans, and support in implementing these solutions.

A key best practice of EcoHouse is to design services based on the needs and perspectives of citizens, to allow an easy understanding of both available options and immediate gains. Another best practice is to nurture close partnerships with the stakeholders (i.e. local network of welfare, housing, education, migrant, and community organizations) as well as city departments. Such partnerships guarantee a diffusion of awareness about the EcoHouse services in the local context, which significantly raises the response rate. Partnerships also facilitate access either to funding schemes for citizens or to job demand from companies. Indeed, a key success indicator of EcoHouse is the support provided to low-income citizens. For instance, low-income households can receive a special zero-interest loan and are supported in negotiating with contractors. Furthermore, through collaboration with the social enterprise Levanto, EcoHouse offers training programs to the unemployed, in order to enable them to apply to calls from local companies on jobs related to building renovation and energy transition. Even though replicated in some cities in Flandres, EcoHouse model may not be easily replicable in other contexts because of the high financial and human resources needed. Reaching the target group of low-income households requires analyzing and integrating their overall needs beyond the issues strictly related to energy. Moreover, the involvement of citizens in the initiative was sponsored by the city authorities through dedicated personnel and specifically designed informative materials, with tips on how to save money through energy-saving behaviors and renovations.

The Energy cooperative **KLIK** (https://netzerocities.app/resource-2619; https:// klikninaodrzivo.com/), offers a similar example from Croatia, which has been active since 2020. KLIK differs from the case of Antwerp as it is a bottom-up initiative: the project was originated by a group of citizens and then supported by the City of Križevci. The cooperative KLIK encourages citizens to invest in renewable energy sources, both in public projects and in their households, and offers support in applying for funds and implementing renewable energy projects. KLIK exemplifies the effectiveness of self-empowerment in the energy transition. Since a group of citizens manages the cooperative, the issues related to lack of trust are significantly reduced, and the needs of users are better understood and addressed than in initiatives managed by public officers alone. The cooperative also managed to engage further institutional stakeholders, and it supports companies and the City of Križevci in writing and submitting projects, in mapping and designing solar power plants, and in organizing further sustainability and engagement activities. KLIK works thus as an intermediary between citizens and other stakeholders in implementing the local availability of expertise and in channeling the funding for the energy transition. A cooperative such as KLIK can be potentially established in every community where there is already an active group of citizens committed to energy transition. Important key elements for the replication of such an initiative are the strong support from local authorities, securing funding for activities and personnel of the cooperative, and the degree of viability of the cooperative model in the national regulatory context.

Since 2019, the **City Council of Valencia** (https://netzerocities.app/resource-3110; https://climaienergia.com/es/oficina/quienes-somos/) has been promoting local energy communities by providing legal advisory and mediation skills to promote agreements among neighbors. The distinctive best practice is the organizational innovation process, which aims at refocusing energy efficiency efforts and aligning citizens on renovation projects. The City Council has developed a template to create the legal profile of energy communities as associations, and it facilitates workshops in each district to design new energy communities in a participatory manner. In parallel, the City Council is also promoting pilots in public and private buildings to test different energy production models. Its initiative also generates a chain effect among neighboring communities asking for city services and accompaniment. The engagement of all citizens is also achieved by the contract between the City Council and the energy communities, which guarantees that the former will cover the investment needed by low-income citizens, lifting them from the initial cost.

There are a couple of interesting takeaways from this initiative. First, it is very important to offer accompaniment, training workshops and legal advisory services at the district level to engage the tenants. The City Council of Valencia exerts such territorial support through the Energy Office, a one-stop-shop for citizens. Secondly, some pilots in public buildings could be first realized to make apparent the advantages and to gain approval from tenants. Such a strategy would overcome possible skepticism by private householders, who may hinder the renovation works. While considering the replication of this experience, essential elements of the initiative are the political commitment from local authorities, the availability of initial funding for pilots and assistance services, and the financial support of low-income communities. Alignment with energy distribution companies is a key pre-requirement to reconfigure the load capacity of the energy system. As an additional step, the City Council is designing a private–public-partnership company to manage the energy communities.

The **SONNET City Lab** (https://sonnet-energy.eu/portfolio-item/bristol/; https://netzerocities.app/resource-2627) in Bristol, U.K., provides some useful case studies on merging engagement and financial strategies (Davis and Cartwright 2019). From January 2020 to October 2021, it experimented with an engagement activity based on the design of a dedicated crowdfunding instrument named Community Municipal Bond (CMB). The Community Municipal Bond allows the municipality to raise capital from citizens, who become direct investors in the energy transition of their city. The city balance sheet guarantees the financial risk and is not transferred to investors. The Community Municipal Bond can thus support or replace the ordinary sources of borrowing of cities to fund specific public infrastructure projects or to refinance existing loans. It is important to stress that Community Municipal Bond is primarily an engagement process rather than just a financial instrument proposed top-down to citizens. The process in Bristol started with engaging building managers and technically surveying the buildings to assess the actual financial investment needed and the potential obstacles and barriers. The second step was surveying citizens to check their interests and identify investment bracket ranges that they would sign up for. Engagement activities are necessary to correctly identify the minimum entry level for the bond investment, which is particularly important to keep lower-income areas engaged. An important element to consider for the replication of the initiative elsewhere is that standard crowdfunding business models must be modified to align with the treasury management protocols of the local authorities. Furthermore, an effort should be made to make easy the understanding and access to Community Municipal Bond for non-financially literate citizens.

The case studies briefly outlined in this section cover some best practices in supporting energy transition in local areas, whose experiences converge on three main issues. First, easy access to services, and not just the delivery of services, is a significant success factor in the engagement of the population involved. Establishing a single touchpoint for citizens has proven very effective since it lifts the burden of the change carried by single households. User-centred design methods are a valuable approach for securing easy accessibility and usability to services (Wever et al. 2008; Wilson et al. 2010). The second issue highlighted is the importance of designing touchpoints and services starting from the needs and the constraints of the target community of citizens, especially considering low-income and vulnerable social groups. Such case studies thus corroborate the adoption of participatory methods for enhancing the transition to new energy systems (Lennon et al. 2019). Eventually, the case studies point out the effectiveness of nurturing networks of collaboration and mutual exchange of information among local public institutions, companies and non-profit organizations. Such networks can mobilize and convey new resources and stipulate win–win agreements without calling for significant financial resources. From this perspective, policymakers and local authorities should either act as inter-mediaries among different local actors or should facilitate the emergence of such intermediaries (Kivimaa et al. 2019).

4 Behavioral Change

This section focuses on three selected case studies of social innovation that under-score effective strategies to foster systemic behavioral change for sustainability. The case studies draw attention to the factors affecting behavioral change including eco-consumption choices and decision making (Panda et al. 2020), personal and social factors (Gifford and Nilsson 2014), and associated cognitive load (Kahneman 2011). In addition, it exemplifies methods such as building awareness, facilitating training, community building, citizen engagement and employing game mechanics towards fostering behavioral change.

Furthermore, the selection touches upon a rich diversity of actors, contexts, and sectors, rendering a comprehensive canvas of Social Innovation in action at various stages (Murray et al. 2010). The case of Climate Meal (Finland) leverages data to foster informed pro-environmental decision making among citizens and local (restaurant) businesses. On the other hand, 1.5-Degree Lifestyle (Finland), harnesses the outreach of smartphone technology to promote sustainable lifestyle choices and consumption patterns. And lasty, the case of PlayUC (Netherlands, Belgium and Austria) takes a gamified approach to engage citizens in pro-environmental activities.

The case of **Climate Meal** (https://netzerocities.app/resource-2847) in the city of Helsinki illustrates how micro changes have positive behavioral impacts. In support of the Mission Zero Foodprint Project (Forum Virium Helsinki 2020), The Climate Meal initiative encourages restaurants (especially ones that operate within the Small and Medium Enterprise category) to offer climate friendly meals in their menu through a campaign that provides them with the Climate Meal label, including tools for calculating the carbon footprint of dishes and tools for communicating their commitment. Restaurants can label items on their menu as "Climate Meals" if the total carbon footprint of all its ingredients does not exceed 1.0 kg CO_2e, which is significantly lower than the Finnish average. Besides the financial potential of serving climate friendly meals to customers, restaurants recognize the need to become more climate neutral and are hence incentivized to make changes to their menus. On the other hand, for customers, the labeling of climate-friendly meals streamlines their pro-environmental decision-making process, making it easy for them to reduce their carbon footprint.

The participating restaurants reported that the demand for vegetarian food increased during the campaign. This showcases the impact of micro level changes towards shaping Pro-Environmental Behavior among both businesses and consumers. The case highlights how social innovation can be focused on lowering the barrier to climate neutrality at the grassroot level. The campaign essentially lowered this barrier especially for restaurants belonging to the SME category that lack the technical know-how or resources to track their carbon footprint. Moreover, it also simplified pro-environmental decision-making for consumers by lowering the associated cognitive load (Kahneman 2011) through an easily identifiable visual marker (i.e., the climate meal label) that distinguishes climate –friendly options, thus, catalyzing behavioral change.

The **1.5 Degree Lifestyles Puzzle** (https://netzerocities.app/resource-3866) is a tool used to help Finnish households and other stakeholders identify lifestyle changes necessary in order to significantly drop their carbon footprint. Based on an online consumption-driven carbon footprint calculator called "Lifestyle Test" set up by the Finnish foundation Sitra in 2017, individuals can calculate their carbon footprint and take significant actions towards climate neutrality. The tool is aimed at supporting personal lifestyle management and leverages the potential of gamification (Bardhan et al. 2015) and widespread outreach (through smartphones) to foster Pro-Environmental Behavior among European citizens. This case particularly showcases the potential of scalable and adaptable technologies for behavioral change.

The need for decarbonized lifestyle and household consumption patterns for limiting global warming to 1.5° as per the Paris Agreement has been echoed extensively by several reports from the Intergovernmental Panel on Climate Change (IPCC) and United Nations Environmental Programme (UNEP) since 2018 (Koide et al. 2021). Direct and indirect emissions from household consumptions comprise a majority of global GHG emissions levels (Hirano et al. 2016), and this statistic is rapidly on the rise. Hence, fostering behavioral change at the grassroot level becomes imperative and inevitable. The 1.5-Degree Lifestyle Puzzle project stands out in this regard as it attempts to foster scalable behavioral change by influencing consumption patterns and sustainable life-style choices directly from the grassroot level through the medium of accessible smartphone technology. Since its initial launch, the app has recorded over a million use sessions, thus indicating the presence of public interest. However, the anticipation of this initial demand diminishing over-time among citizens due to wavering interests and concerns over the lowering quality of life cannot be ruled out. Moreover, the effects of nudging (Lehner et al. 2016) or incentives presented by a smartphone application over personal factors towards consumption and lifestyle choices such as pricing, convenience or associated cultural influences (Gifford and Nilsson 2014) remains an open question for further evaluation.

PlayUC (https://netzerocities.app/resource-3923), Netherlands, Belgium and Austria highlight the use of co-located serious games as a means to raise awareness and engage (primarily young) adults to reduce their carbon footprints and promote climate-friendly decision-making. The intension behind the use of serious games, both in digital and physical forms, is to fulfil the functional scope of educating or training citizens to reduce their carbon footprints. The aim of this research-based initiative is to foster a comprehensive understanding of complex urban problems through story-telling and gamified participatory processes. The initiative involves various innovative game prototypes encompassing different thematic areas of carbon footprint aimed at target groups, from mobile applications simulating traffic patterns to board games fostering urban development campaigns. The games are contextualized for the participants to simulate realistic and relatable scenarios of urban challenges in their cities in order to raise awareness or motivate participants to act.

The growing body of research on serious games towards sustainability (Stanitsas et al. 2019) highlights its potential of engaging citizens to collaborate and foster systemic behavioral changes, both as a collective or at an individual level. PlayUC has particularly pointed out the need for such self-organization of citizens, which can be a powerful tool to promote Pro-Environmental Behavior (PEB) in cities.

These three case studies exemplify effective Social Innovation strategies for driving systemic behavioral change toward sustainability, and present key factors that drive the change. In the first case of Climate Meal, it's noteworthy to consider the aspect of reducing cognitive load (Sweller 2010) towards climate-friendly decision-making. In the second case, outreach and accessibility are also important considerations towards fostering behavioral change at scale rapidly. Finally, the role of gamification (Douglas and Brauer 2021) as seen in the case of PlayUC underscores the need for relatable story-telling and engaging experiences to facilitate the pro-active participation of citizens. While there is a multitude of factors affecting systemic behavioral change, these case studies provide valuable insights from the practical application of Social Innovation and highlight effective means to navigate the complexities of bottom-up approaches towards fostering systemic behavioral change.

5 Implications and Conclusion

The cases presented in this chapter provide cities' transition teams, public administrators and urban stakeholders with evidence of the relevance of deploying social innovation for reducing GHG emissions and simultaneously improving social impact at urban level. Beyond—and in addition to—technological innovations, the crucial contribution of innovations in social practices for cities' sustainability are increasingly recognized both at theoretical level (Ceschin and Gaziulusoy 2016; Diepenmaat et al. 2020; Bresciani et al. 2022) and by public administrations and politicians, as outline by the above presented cases. A healthy critical perspective can support the analysis and potential replication of such cases: can the success of these cases be replicated in different contexts? Under which conditions? Which resources are needed? The answer to these questions requires a much more extended explanation: while the purpose of this chapter has been to illustrate a breadth of cases clustered into key categories, an extended illustration of each case is available on the NZC platform, together with thirty addition in-depth case studies of social innovation for climate neutrality (https://netzerocities.app/resource-4074). The next chapter provides an illustrative in-depth case study with a portfolio of activities.

References

Anttiroiko A (2016) City-as-a-platform: the rise of participatory innovation platforms in finnish cities. Sustainability 8(9), article 922. https://doi.org/10.3390/su8090922

Bardhan R, Bahuman C, Pathan I, Ramamritham K (2015) Designing a game based persuasive technology to promote pro-environmental behaviour (PEB). In: 2015 IEEE Region 10 humanitarian technology conference (R10-HTC). https://doi.org/10.1109/r10-htc.2015.7391844

Bresciani S, Rizzo F, Deserti A (2022) Toward a comprehensive framework of social innovation for climate neutrality: a systematic literature review from business/production, public policy, environmental sciences, energy, sustainability and related fields. Sustainability 14(21):13793

Ceschin F, Gaziulusoy I (2016) Evolution of design for sustainability: from product design to design for system innovations and transitions. Des Stud 47:118–163

Davis M, Cartwright L (2019) Financing for society: assessing the suitability of crowdfunding for the public sector. University of Leeds. https://core.ac.uk/download/pdf/199226065.pdf. Last accessed 2023/10/14

Diepenmaat H, Kemp R, Velter M (2020) Why sustainable development requires societal innovation and cannot be achieved without this. Sustainability 12(3):1270

Douglas BD, Brauer M (2021) Gamification to prevent climate change: a review of games and apps for sustainability. Curr Opin Psychol 42:89–94. https://doi.org/10.1016/j.copsyc.2021.04.008

Eckhardt J, Kaletka C, Krüger D, Maldonado-Mariscal K, Schulz AC (2021) Ecosystems of co-creation. Front Sociol 26(6), article 642289. https://doi.org/10.3389/fsoc.2021.642289

Forum Virium Helsinki (2020) Mission zero foodprint—Forum Virium Helsinki %. Forum Virium Helsinki. https://forumvirium.fi/projektit/mission-zero-foodprint-hanke/

Galego D, Moulaert F, Brans M, Santinha G (2022) Social innovation & governance: a scoping review. Innov: Eur J Soc Sci Res 35(2):265–290. https://doi.org/10.1080/13511610.2021.187 9630

Gifford R, Nilsson A (2014) Personal and social factors that influence pro-environmental concern and behaviour: a review. Int J Psychol 49(3):n/a-n/a. https://doi.org/10.1002/ijop.12034

Hirano Y, Ihara T, Yoshida Y (2016) Estimating residential CO_2 emissions based on daily activities and consideration of methods to reduce emissions. Build Environ 103:1–8. https://doi.org/10.1016/j.buildenv.2016.02.021

Kahneman D (2011) Thinking, fast and slow. Doubleday Canada

Kivimaa P, Boon W, Hyysalo S, Klerkx L (2019) Towards a typology of intermediaries in sustainability transitions: a systematic review and a research agenda. Res Policy 48(4):1062–1075

Koide R, Lettenmeier M, Akenji L, Toivio V, Amellina A, Khodke A, Watabe A, Kojima S (2021) Lifestyle carbon footprints and changes in lifestyles to limit global warming to 1.5 °C, and ways forward for related research. Sustain Sci 16(6):2087–2099. https://doi.org/10.1007/s11625-021-01018-6

Lehner M, Mont O, Heiskanen E (2016) Nudging—a promising tool for sustainable consumption behaviour? J Clean Prod 134:166–177. https://doi.org/10.1016/j.jclepro.2015.11.086

Lennon B, Dunphy NP, Sanvicente E (2019) Community acceptability and the energy transition: a citizens' perspective. Energy, Sustain Soc 9(1):35

Murray R, Caulier-Grice J, Mulgan G (2010) The open book of social innovation. The Young Foundation

Panda TK, Kumar A, Jakhar S, Luthra S, Garza-Reyes JA, Kazancoglu I, Nayak SS (2020) Social and environmental sustainability model on consumers' altruism, green purchase intention, green brand loyalty and evangelism. J Clean Prod 243(1):118575. https://doi.org/10.1016/j.jclepro.2019.118575

Stanitsas M, Kirytopoulos K, Vareilles E (2019) Facilitating sustainability transition through serious games: a systematic literature review. J Clean Prod 208(1):924–936. https://doi.org/10.1016/j.jclepro.2018.10.157

Sweller J (2010) Element interactivity and intrinsic, extraneous, and germane cognitive load. Educ Psychol Rev 22(2):123–138. https://doi.org/10.1007/s10648-010-9128-5

Terstriep J, Rehfeld D, Kleverbeck M (2020) Favourable social innovation ecosystem(s)?—An explorative approach. Eur Plan Stud 28:1–25. https://doi.org/10.1080/09654313.2019.1708868

The GovLab (2023) SynAthina. Blog post by Equitable Engagement Lab at the Burnes Center for Social Change. https://equitableengagementlab.org/synathina. Last accessed 2023/10/16

Wever R, Van Kuijk J, Boks C (2008) User-centred design for sustainable behaviour. Int J Sustain Eng 1(1):9–20

Wilson G, Bhamra T, Lilley D (2010) Reducing domestic energy consumption: a user-centred design approach. https://repository.lboro.ac.uk/articles/Reducing_domestic_energy_consumption_a_user-centred_design_approach/9341027/files/16949720.pdf. Last accessed 2023/10/14

Activating Ecosystems for Change by Enriching the Civic "Soil" for Social Innovation. The Wiener Klimateam Project as a Case Study

Daniela Amann⑩, Arild Ohren⑩, Max Stearns⑩, and Ilaria Mariani⑩

Abstract Through the analytical exploration of the 'Wiener Klimateam' case, the chapter exemplifies the pivotal role of a sound democratic ecosystem to activate social innovation (SI) and drive urban transformation towards climate neutrality. It shows the relevance of integrating diverse stakeholder perspectives to address complex climate challenges, leveraging citizen engagement and democratic processes. This case study shows how comprehensive analysis, stakeholder engagement, and participatory design can catalyse systemic changes within city infrastructures, emphasising inclusivity and equitable change. The initiative's innovative approach, including a citizens' jury, showcases the transformative potential of a thriving democratic ecosystem and SI in operationalising sustainable urban governance and fostering collaborative, cross-sectoral strategies. By doing that, the chapter aims to derive insights that are potentially scalable and extendable, focusing on the dynamics of urban innovation. It will examine how democratic ecosystems have been established and SI operationalised to address the challenges that cities face on their journey towards climate neutrality, bridging the gap between theoretical discourse and practical application. The analysis is presented against the dimensions of theoretical constructs derived from Chaps. 2 and 3 and summarised in a framework to explore the operational challenges and opportunities in real-world contexts.

Keywords Deliberative decision-making · Citizen engagement · Democratic infrastructure · Case study

D. Amann · A. Ohren · M. Stearns
Democratic Society, Brussels, Belgium

I. Mariani (✉)
Department of Design, Politecnico di Milano, Milan, Italy
e-mail: ilaria1.mariani@polimi.it

© The Author(s) 2025
S. Bresciani (ed.), *Social Innovation Projects for Climate Neutral Cities*,
PoliMI SpringerBriefs, https://doi.org/10.1007/978-3-031-87726-1_6

1 Introduction[1]

Current technological advancements, market-driven strategies, and regulatory measures have fallen short in addressing critical ecological issues such as climate change and biodiversity loss, highlighting their limitations in resolving these human-induced environmental challenges (Baste et al. 2021; Lee et al. 2024). This situation underscores the urgency for profound societal transformation (Haskell et al. 2021), which entails reevaluating and altering existing social practices, organisational structures, and prevailing norms and beliefs (Heikkurinen et al. 2016). Such a shift aligns with the principles of a robust sustainability model, advocating for deeper, systemic changes (Ekins et al. 2003; Stål and Bonnedahl 2016). In this critical juncture, inclusive, just, and effective social innovation (SI) emerges as a key driver of transformative change (Avelino et al. 2017), playing a critical role in disrupting and reshaping entrenched institutions within specific societal contexts and redefining social relationships (Pel et al. 2020).

European cities stand at a crossroads of transformation, where the adoption of a more integrated approach to innovation becomes necessary, synergising social challenges and technical advancements (Appio et al. 2019). Yet, as emerged from previous discussion, the transition towards climate neutrality is laden with multi-dimensional complexities, including entrenched bureaucratic inertia and the need for a paradigm shift in public administration towards more inclusive and innovative practices (Howaldt et al. 2019; Wittmayer et al. 2020). Such challenges are so deeply interwoven within the institutional, industrial, and societal sectors to require a systemic approach to transformation. Comprehensive frameworks established in earlier discourses on urban systems and emissions (Kılkış 2022; Lwasa and Seto 2023; Nagesh et al. 2021) and SI as a transformative catalyst (Avelino et al. 2017; Dias and Partidário 2019; Mehmood et al. 2020) are fundamental in deconstructing and unpacking the existing complexity, providing foundational understanding for a strategic reorientation towards collaborative and innovative urban governance. Cities are indeed recognised as hubs where a diverse array of actors are constantly inspired to engage in innovations tailored to meet specific needs, whether they pertain to the market, organisational dynamics, or community interests (Concilio et al. 2019).

Traditionally, urban interventions oriented to climate neutrality have often relied on top-down approaches or continued to employ established, yet possibly outdated models. While these models may be familiar, they can unintentionally intensify existing urban challenges. This insight prompts the exploration of the 'Wiener Klimateam' project as a case study, showcasing how Vienna has tackled the issue

---

[1] This chapter draws on two complementary streams of expertise: Daniela Amann, Arild Ohren, and Max Stearns from Democratic Society (Demsoc) are practitioners in participation and democratic innovation, while Ilaria Mariani offers a scholarly perspective at the intersection of Design for Public Sector Innovation and Social Innovation. Together, these viewpoints provide a holistic and comprehensive approach to the chapter and the case study it explores, emphasising the role of democratic innovation in facilitating social innovation.

adopting a mixed approach that blends a bottom-up approach with participatory practices that foster collaboration, shared decision-making, and empowerment. While organised by the city to involve citizens, this approach demonstrates that well-designed governmental initiatives can effectively engage public participation and contribute to democratic processes. Specifically, the case will demonstrate how the effective introduction of a democratic ecosystem and infrastructure not only fosters but also amplifies SI, emphasising the critical role of active engagement as an essential precondition for catalysing SI within the city. This case addresses some of the shortcomings of traditional public sector innovation methods (Wittmayer et al. 2020) discussed in the current scientific discourse, advocating for effective civic engagement and innovation competitions as strategies to elicit grassroots change and give a push to the urban innovation landscape (Hartmann et al. 2019). This trajectory also meets an ongoing reflection on how responsible transformation should refrain from the pitfalls of rapid, undemocratic innovation (Irani 2019), emphasising the importance of thoughtful, democratic deliberation. This mirrors Gordon and Walter's (2019) notion of 'meaningful inefficiency' that champions innovation as reflective and inclusive. A slower pace of innovation creates a fertile ground for transformative interactions, fostering a rich tapestry of community ideas and collaborations (Gordon and Mugar 2020). This pace fosters a productive condition and setting for encounters where the exchange of lived experiences, imaginaries, and possible futures can meaningfully intertwine to inspire better actions.

Overall, the scholarly discourse delineates the challenges and opportunities presented by SI, advocating for a strategic integration of SI principles into urban governance to overcome barriers and leverage the potential for impactful climate action (Hoppe and De Vries 2019; Reynolds et al. 2017).

This chapter leverages critical thinking from across these fields, as well as insights from The Democratic Society's (Demsoc) fifteen years of practice to discuss an instance of a novel approach for the establishment, maintenance, and reimagination of democratic infrastructures capable of affording, accelerating, and sustaining impactful social innovations (https://www.fdsd.org/democratic-infrastructure/). Informed by a long experience supporting governments and communities across Europe to foster more participatory, deliberative, and co-creative decisions, democratic infrastructure is here used to refer to persistent sets of socio-political, socio-cultural, and socio-technological elements that enable the continuous development and reinforcement of democracy in a specific setting. This entails a democracy-oriented plan tailored to the local context, the necessary capabilities, networks, and governance mechanisms in a setting that embodies flexibility, forethought for long-term impacts, and the ability to foster democratic self-enhancement.

As a systemic problem, climate change is a major challenge for traditional representative democracies and city administrations. Climate action often involves decisions with a great political and financial risk that require trade-offs, cross-silo collaboration, as well as projects and strategies that go beyond electoral cycles. It often involves allegedly unpopular political decisions, as for example, moving from a car-focused urban planning to a more walking and cycling friendly. Moreover, climate change affects social groups very differently, with the vulnerable, the elderly and the sick suffering particularly from its effects. As in many major cities, also in Vienna, it poses a particular challenge, as many of these groups are among the traditionally underrepresented groups in political processes and often go unheard in decision-making around climate action. Thus, the research questions that emerge from this discourse are: How can cities operationalise inclusive and just SI to encourage a responsible and desirable pathway towards climate neutrality? How can a democratic ecosystem support cities in ensuring equitable and effective SI practices to create inclusive and just climate action?

To answer them while investigating the potential link between democratic ecosystems as prerequisite infrastructures designed to enable and enhance inclusive, just, and effective SI in decision-making, this chapter focuses on the project 'Wiener Klimateam' (Vienna Climate team), undertaken by Demsoc in partnership with the City of Vienna from 2020 to 2022 as a case study (Stake 1978, 1995; Yin 1984). It specifically investigates the activation of a consistent ecosystem founded on the active engagement of citizens and stakeholders, as a strategy employed to actualize the theoretical potential of SI. Therefore, the chapter discusses a case of how cities can value and integrate a broad spectrum of civic voices and experiences to progress towards climate neutrality, thus underscoring the significance of participatory governance as a cornerstone for advancing towards climate neutrality. Demsoc supported the city in the conceptual phase (2020), the co-design phase that involved city staff from all relevant departments (2021), as well as the implementation phase, where it was responsible to design and implement the deliberative citizens jury (2022).

2 Four Dimensions of Observation

As discussed in Chaps. 2 and 3, cities are in need of an approach to societal challenges that considers human, social, ecological, and economic dimensions (Grimm et al. 2013; Hoppe and De Vries 2019; Hubert 2010). In this context, the capacity to initiate and sustain SI processes that transform beyond their current paradigms and structures emerges as a critical determinant of successful urban transformation. The spur for change often arises from the rich soil of community engagement and co-creation, fostering a fertile ground for the seeds of innovative thought and action to grow (Amatullo et al. 2021)—in this context, democratic ecosystems can be regarded as an instance of how such engagement and co-creation manifest, facilitating a structured yet flexible environment where innovation is nurtured and sustained.

It advocates for integrating these innovations into broader, strategic portfolios, emphasising their intrinsic value beyond mere supplementation to technical solutions (Wittmayer et al. 2020). However, it is recognised that a constant challenge in embedding such innovative practices within public administrations, often lies in entrenched bureaucratic procedures (Reynolds et al. 2017). The discourse consistently underscores the necessity of institutionalising SI including practices such as co-creation and participatory governance to foster a comprehensive pathway towards climate neutrality. The transformational potential of initiatives such as the RATKAISU 100 (https://www.sitra.fi/en/topics/ratkaisu100) exemplifies the contribution of inclusive competitions in bringing innovative solutions to the forefront. Such approaches align with the need to encourage broad(er) participation to nurture creative responses to pressing societal issues. Embedding transformative SI within institutional governance presents significant challenges, particularly concerning dynamics and limitations of transition-oriented analyses during crucial periods of institutional transformation (Bauler et al. 2017; Shabb and McCormick 2023). Overall, the landscape underscores the necessity for a framework that not only bridges insights from experimentation but also acknowledges past failures and leverages the benefits of active community involvement to guide sustainable transformations. Moreover, adopting open and participatory frameworks in policy design, as highlighted by Reynolds and colleagues (2017), cannot be overstated in the context of achieving systemic and sustainable urban transformation. This chapter explores the practical integration of these adaptive, participatory, and iterative methodologies into urban governance, aiming at providing insights to cities in advancing towards a transition to climate neutrality that balances ecological needs with social equity (European Commission 2019).

This chapter recognises these contemporary challenges and seeks to tackle them by exploring how theoretical insights can be effectively translated into real-world applications. In light of this, the discourse unfolds against the theoretical underpinnings outlined in previous chapters, and summarised in Table 1, emphasising the role of democratic ecosystems in fostering engagement and enabling social innovation within urban settings.

Drawing on this comprehensive theoretical framework, this chapter delves into the practical application of these concepts, creating a bridge between theoretical understanding and empirical evidence gathered from past experimental efforts. Through the analytical exploration of the 'Wiener Klimateam' case, it aims to derive potentially scalable and extendable insights, focusing on the dynamics of urban innovation. The analysis is presented against the four dimensions summarised in the framework above, and explores the operational challenges and opportunities in real-world contexts, demonstrating how democratic ecosystems can serve as foundational structures for the effective operationalisation of social innovation in urban contexts.

3 The Wiener Klimateam Project: A Case Study

The Wiener Klimateam is a governance and participation innovation project to promote climate change mitigation and adaptation projects for a climate-neutral and liveable Vienna. Initiated under the EIT Climate-KIC Deep Demonstrations project in 2020 (https://www.climate-kic.org/wp-content/uploads/2019/10/Healthy-Clean-Cities.pdf), it aims to tackle key barriers for Vienna's goal to become climate neutral by building the project upon three systemic levers: complementing representative democracy with participatory elements, promoting cross-silo collaboration, and hereby creating the foundation for ambitious climate action.

Table 1 Body of theoretical knowledge organised into four clusters

Clusters	Dimensions	Relevant literature	Chap. ref.
1. Role of cities in climate neutrality	Cities as both sources of emissions and hubs of innovation and action for climate change mitigation	Recognition of the dual role of cities: Lwasa and Seto (2023) Cities as hubs for innovation: Concilio et al. (2019)	Chap. 2
	European policy context: European Green Deal and the 100 Climate-neutral and Smart Cities Mission as driving forces	European Green Deal and Cities Mission: European Commission (2019)	Chap. 3
2. SI as a transformative force	SI for climate neutrality, focusing on what works for cities and citizens	Importance of SI for climate neutrality: Hoppe and De Vries (2019) Broad array of SI practices: Wittmayer et al. (2020)	Chap. 2
	SI role for just and sustainable transitions, to be embedded in policy and practice	SI definition and importance: Hubert (2010) SI as a transformative capacity: Grimm et al. (2013)	Chap. 3
3. Operationalising SI in urban governance	Roles of cities in SI: Enablers, Mediators, and Visionaries	Cities as Enablers, Mediators, Visionaries: Demos Helsinki's work in the Helsinki City Region (Schmidt-Thome and Mesarcik, Ch. 3)	Chap. 2
	Approaches to adoption, from SI as a policy outcome and integrating it into policy-making	SI as policy output and approach: Reynolds et al. SI in the policy discourse, from theory to practice: Grimm et al. (2013)	Chap. 3
	Transition team concept and role in orchestrating city-level actions for climate neutrality	Transition Teams in orchestrating actions: Shabb and McCormick (2023)	Chap. 3
4. Challenges and strategies for SI implementation	Barriers to SI adoption in public administrations	Difficulties in public administrations: Howaldt et al. (2019)	Chap. 3
	Difficulties cities face in promoting SI and the opportunities it presents for climate action	Instrumentalisation of SI: Wittmayer et al. (2020) SI in energy cooperatives and participatory governance: Hoppe and De Vries (2019)	Chap. 2
	Civic engagement and competition in validating and legitimising SI	Importance of civic engagement: Hartmann et al.	Chap. 2

Table 2 Levers to address challenges

Lever	Directions
Engaging citizens into decision-making processes	Catalyse climate actions by facilitating collective learning among citizens, city experts, and politicians, thereby fostering a culture of democracy both within the city administration and the broader community
Complementing representative democracy with participatory elements	Enhance democratic participation, augmenting the existing representative democracy framework with participatory elements and ensuring the inclusion of traditionally underrepresented groups. Notably, the project is meant to enable Viennese residents lacking formal voting rights to participate, thereby broadening the scope of democratic engagement and decision-making
Transcending a siloed approach	Act as a catalyst for organisational change, specifically by encouraging cross-departmental collaboration and policy prioritisation by citizens. This strategy aims to overcome traditional siloed approaches within the city administration, promoting a more integrated and cohesive approach to climate action and governance

Project inception. The idea for the Wiener Klimateam was developed in the framework of the EIT Climate-KIC Deep Demonstrations project in 2020 (https://www.climate-kic.org/wp-content/uploads/2019/10/Healthy-Clean-Cities.pdf). The goal of this project was to design portfolios of innovations capable of unlocking transformation across city systems. During this process, the City of Vienna, together with a consortium of experts in the fields of citizen engagement, democratic decision-making, governance, social innovation, economic analysis, policy development, and scaling identified key barriers for reaching climate neutrality. The Wiener Klimateam was identified as a possible key cross-cutting enabler of three of the main challenges Vienna was facing at the time: the activation and involvement of citizens in decision-making processes, innovating governance processes and launching major climate projects across silos with citizens' support.

Comprehensive analysis and stakeholder consultation. Following the initial idea, a comprehensive analysis and dialogues with a diverse array of experts was conducted. Initial efforts included a thorough examination of Vienna's existing participatory budgeting processes, including case studies analysis and expert interviews (n.10), alongside reports evaluating participatory budgeting case studies globally (n.8). Through extensive consultations at all levels of governance, the project gains endorsement (http://www.wien.gv.at/regierungsabkommen2020/files/Koalitionsabkommen_Master_FINAL.pdf)[2] while being recognised as a strategic enabler for overcoming critical challenges, summarised in Table 2.

The comprehensive analysis ran to inform the project underpins a design strategy keenly responsive to the diverse needs of stakeholders, showcasing a commitment

---

[2] The comprehensive report detailing the procedures, approval, and budgetary allocation associated with the activity is documented in German.

to build upon established practices and case studies but also with a broad spectrum of experts to enrich the process at multiple levels. Particularly, the identification of challenges and strategic enablers highlights a grasp of harnessing citizen engagement in decision-making processes, advocating for a collective approach to climate action that breaks down traditional departmental barriers. This strategy clarifies the importance of democratic ecosystems in facilitating such engagement and innovation.

The challenges and strategic enablers portray a clear approach to leveraging citizen engagement for making climate action initiatives more grounded in the community's actual needs and aspirations, ultimately leading to more targeted and effective solutions. This innovation in governance not only enhances the efficiency and desirability of the solution proposed, but also contributes to building trust, credibility, and sense of ownership among stakeholders. Embedding participatory elements into the decision-making framework serves to render governance more inclusive, transparent, and reflective of the collective will, thereby enhancing overall legitimacy. Furthermore, overcoming a siloed approach can lead to more integrated and potentially innovative strategies that build upon the strengths and resources of various departments and sectors, encouraging the sharing of insights and best practices. Additionally, involving citizens throughout the process ensures that initiatives garner widespread support, augmenting their feasibility and impact in tackling the complex challenges of climate change.

Co-design process and final design. In the latter part of 2020 and with the support of Demsoc, the City of Vienna initiates the project design phase as a co-design process. This endeavour aims to actively involve a broad spectrum of stakeholders, including representatives from both the city and district administrations, civil society, and political leaders. The process is structured around five workshops strategically organised to achieve three primary objectives: firstly, to holistically incorporate the diverse needs and viewpoints of city and district administrations into the project's framework; secondly, to develop and improve participatory practice skills among city departments in preparation for implementation; and thirdly, to create a cooperative environment that encourages cross-departmental collaboration for more cohesive climate action and governance. This process underscores the role of democratic ecosystems in fostering a culture of engagement and innovation, setting the stage for social innovation through inclusive and effective stakeholder participation.

In addition, an advisory committee composed of representatives from civil society representatives is formed to offer critical insights into the project's conceptual underpinnings, enhancing its ability to address complex challenges. This co-design process exemplifies a foundational principle of SI, emphasising effective participatory engagement and collaborative design as essential mechanisms for fostering sustainable change. By integrating diverse stakeholder perspectives, the process ensures that solutions are not only inclusive but also reflective of the diverse needs and aspirations of the community—namely enhancing the design of innovative, community-driven solutions.

By the close of 2021, a definitive design for the Wiener Klimateam is established, garnering the endorsement of involved city departments, political figures, and civil society organisations. The project is inaugurated with a two-year pilot phase,

Fig. 1 Piloting methodology

selecting three districts annually as part of a strategic plan to eventually extend the initiative across the entire city. This phased approach is strategically chosen to evaluate the project's efficacy and scalability within the context of Vienna's broader objectives for climate neutrality and civic participation, including the needs of different social groups in the loop.

First pilot with implementation of the citizens jury. In 2022, the Wiener Klimateam project transitions into its pilot phase, implementing a participatory budget with innovative and deliberative components, such as a citizens jury that empowers community members to directly influence decision-making, ensuring that selected projects address local needs and priorities. Figure 1 outlines the pilot methodology.

While each step included innovative aspects which could be explored, the discussion that follows will focus specifically on the citizens' jury, exploiting Demsoc's critical role in its design and execution. The jury is asked to identify the most impactful projects for their district, fostering an environment for meaningful discussion on multilayered community needs, expectations, and aspirations. As such, it goes beyond its democratic and deliberative function: while only advisory, its recommendations are typically upheld by city and district councils, reflecting a consensus agreement to execute all chosen projects. Figure 2 points out the innovative elements in the jury design, to foster inclusive and deliberative decision-making.

Preliminary outcomes and implications. Early insights into the Wiener Klimateam's two-year pilot, still under evaluation in early 2024, reveal significant progress. The initiative has secured €12 million for climate action in six districts, creating four new municipal roles dedicated to drive climate action, enhancing citizen participation and capacity building. This strategic investment underscores the Wiener Klimateam's crucial role in Vienna's environmental strategy and its commitment to sustainable and inclusive urban development. The first year generated over 1000 ideas, with 200 participants engaged in co-creation workshops. This process resulted in the selection of 19 projects from 102 proposals by 52 citizens representing a cross-section of the city's population, with a total budget of €6.6 million allocated for implementation, spanning from urban greening, cycle path enhancements, and educational trails. The second year sees an increase in submissions with over 1300 ideas, indicating growing community engagement and input into Vienna's climate neutrality goals. Ultimately, Vienna's selection as the European Capital of

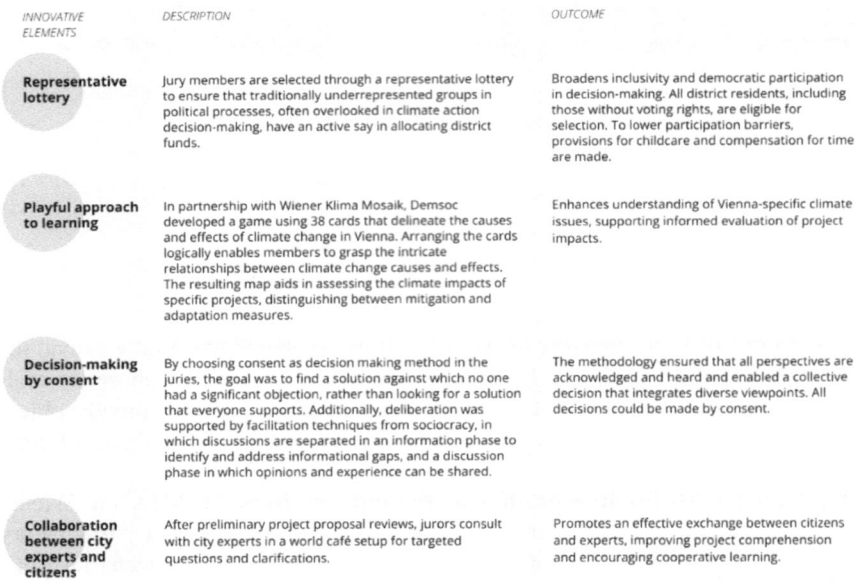

INNOVATIVE ELEMENTS	DESCRIPTION	OUTCOME
Representative lottery	Jury members are selected through a representative lottery to ensure that traditionally underrepresented groups in political processes, often overlooked in climate action decision-making, have an active say in allocating district funds.	Broadens inclusivity and democratic participation in decision-making. All district residents, including those without voting rights, are eligible for selection. To lower participation barriers, provisions for childcare and compensation for time are made.
Playful approach to learning	In partnership with Wiener Klima Mosaik, Demsoc developed a game using 38 cards that delineate the causes and effects of climate change in Vienna. Arranging the cards logically enables members to grasp the intricate relationships between climate change causes and effects. The resulting map aids in assessing the climate impacts of specific projects, distinguishing between mitigation and adaptation measures.	Enhances understanding of Vienna-specific climate issues, supporting informed evaluation of project impacts.
Decision-making by consent	By choosing consent as decision making method in the juries, the goal was to find a solution against which no one had a significant objection, rather than looking for a solution that everyone supports. Additionally, deliberation was supported by facilitation techniques from sociocracy, in which discussions are separated in an information phase to identify and address informational gaps, and a discussion phase in which opinions and experience can be shared.	The methodology ensured that all perspectives are acknowledged and heard and enabled a collective decision that integrates diverse viewpoints. All decisions could be made by consent.
Collaboration between city experts and citizens	After preliminary project proposal reviews, jurors consult with city experts in a world café setup for targeted questions and clarifications.	Promotes an effective exchange between citizens and experts, improving project comprehension and encouraging cooperative learning.

Fig. 2 Jury design innovative elements and their outcome

Democracy for 2024/25 (https://www.wien.gv.at/english/politics-administration/eur opean-capital-of-democracy.html) underscores the city's commitment to enhancing democratic engagement, leveraging insights from the Wiener Klimateam project.

Democratic innovations and inclusion strategies. The Wiener Klimateam project introduced a representative lottery procedure for the Citizens' Jury selection, ensuring that a broad spectrum of residents, including those without voting rights, actively participate in decision-making. This method not only democratises the process but also enriches it with a multitude of perspectives, leading to more nuanced and empathetically scoped solutions. This selection method is of particular relevance for Vienna, as approximately one-third of Vienna's population is not eligible to vote due to non-citizenship and another third' abstention from voting in the 2020 city council elections (voter turnout: 65.27%). Many of those who fall into one of these two groups are among the socio-economically disadvantaged in Austria; they tend to work in professions and conditions that are prone to shifts and therefore tend to be hit harder by systems crises—such as the coronavirus or climate change (Ehs and Zandonella 2021). Within the project, the inclusion strategy included a close cooperation with those most proximate to the target groups. A toolbox was created for these individuals to aid in generating ideas and raising awareness about climate mitigation and adaptation measures among target groups. Feedback from 50 experts across 20 departments of the City of Vienna indicates the initiative successfully broadened civic process inclusion at all levels, with constructive early exchanges and dialogue with citizens.

Institutionalisation and future directions. Vienna's commitment to institutionalising the project underscores its dedication to enhancing participatory governance and ensuring the sustainability of innovative civic engagement strategies. European experts and citizens have chosen Vienna as the European Capital of Democracy 2024/25, showing the city's commitment to continue fostering citizen co-creation with transparent and structured governance. This institutionalisation aims to generate long-term effects in city administration by (i) establishing processes for faster action, decision making and implementations, and (ii) further expanding participation knowledge through increased interdepartmental communication and cooperation. The overarching goal is to transfer insights and methodologies to other projects, promising to solidify the foundation for more integrated, co-creative, and deliberative processes within urban governance.

4 Discussing Findings Against the Theoretical Background

Discussing findings against literature, the Wiener Klimateam initiative emerges as a key example of how deeply understood social needs can act as a catalyst for impactful social innovation. The initiative's foundation recognizes the varied impacts of climate change on different societal groups, shaping its design and results significantly.

Role of cities in climate neutrality. The Wiener Klimateam initiative showcases how cities can serve as catalysts for climate action, acknowledging the differential impact of climate change on various social groups. By centering a democratic approach that involves diverse communities, the project ensures actions are equitable, relevant, and desirable. Engaging experts from various fields, it ensures to be grounded in the latest research and practices, becoming a best practice of how urban areas can lead in achieving climate neutrality. The Wiener Klimateam case study largely exemplifies how cities function as settings of change and innovation hubs, directly engaging with the EU policy context of climate action. The case reports on how a democratic ecosystem facilitates SI through effective multi-stakeholder engagement. It illustrates the proactive stance cities should take, aligning with the European Green Deal and the 100 Climate-neutral and Smart Cities Mission, to mitigate climate impacts through localised innovation and governance strategies (Concilio et al. 2019; Lwasa and Seto 2023).

SI as a transformative force. The case highlights the SI practical applications in addressing urban climate challenges, reflecting on the broad array of SI practices that contribute to a just and sustainable transition. The project's design and implementation process provides tangible examples of how robust and inclusive democratic ecosystems enrich SI as a catalyst for change, particularly through the deliberative citizens' jury, which embodies SI transformative potential. It serves indeed as a model for deepening the understanding of social needs through citizen engagement. The innovative approach of a representative lottery and inclusive jury composition democratises the process, allowing for a wide range of community voices to shape climate action decisions (Hoppe and De Vries 2019; Wittmayer et al. 2020). By involving experts in the conceptual phase, the project shows how a transformative SI initiative can embed sustainable transitions in policy and practice (Grimm et al. 2013). These dialogues ensure that the project is designed to challenge and change dominant institutions and social relations towards more sustainable futures. These dialogues ensure that the project is designed to challenge and change dominant institutions and social relations towards more sustainable futures, facilitated by empowering a democratic ecosystem.

Operationalising SI in urban governance. The case underscores the importance of a multidisciplinary team that supports orchestrating city-level actions towards climate neutrality (Shabb and McCormick 2023). Supporting the concept of cities as enablers, mediators, and visionaries in SI. The case specifically showcases how cities can adopt innovative democratic innovations and SI strategies as policy outcomes and integrate them into policymaking (Grimm et al. 2013; Reynolds et al. 2017). The initial consultations and analyses establish a foundation for cross-silo collaboration and participatory governance, addressing the operational challenges and methodologies for integrating SI within urban frameworks through the activation of a democratic ecosystem. An attitude further demonstrated by the consistent engagement of citizens and city experts in co-creation sessions, while maintaining open dialogues with politicians to ensure governance alignment and support. These methodologies enhance stakeholder engagement and facilitate the scoping, development, and implementation of innovative strategies. Ultimately, the integrated and collaborative approach, combining a bottom-up approach with participatory strategies, exemplifies democratic innovation conducive to continuous experimentation and learning.

Challenges and strategies for SI implementation. The case underscores the necessity of developing key capabilities for robust innovation portfolios and highlights the significance of civic engagement, participation, and democracy—more generally—in enabling, legitimising, and validating SI efforts. The early focus on engaging a broad range of stakeholders directly addresses the barriers to SI adoption within public administrations and the challenges cities face in promoting SI (Howaldt et al. 2019; Wittmayer et al. 2020). These discussions are pivotal in identifying and strategising around the complexities of SI implementation, ensuring the project is designed with an awareness of the institutional and societal hurdles it needs to overcome. The intention to institutionalise the Wiener Klimateam project reflects a strategic response to overcoming barriers to SI adoption. With the intention to standardise processes and transfer knowledge across municipal departments, the project lays the groundwork for more co-creative and deliberative processes in future urban governance, addressing the identified challenges and leveraging opportunities for impactful climate action. This highlights the strategic intention towards creating an ecosystem where democratic practices are not just occasional occurrences but are ingrained in the governance fabric of the city.

5 Takeaways from the Case Study Analysis

This case study sheds light on three principal findings, underscoring how the development and sustenance of democratic infrastructure significantly bolster a city's capability for social innovation. First, the foundation of impactful social innovation processes and outputs lies in a clear understanding of social needs. Engaging citizens actively contributes to identifying, defining, and refining these needs, ensuring that innovation efforts are grounded in real-world contexts and challenges faced by communities. Second, effective scoping, development, and implementation of social innovation require the involvement of a diverse range of stakeholders. Engaging citizens in this process fosters the creation of networks, relationships, and trust—both social and emotional—that are essential for carrying out innovative experiments and scaling them effectively. Third, to ensure that social innovation processes and outcomes go beyond isolated interventions or singular events, there must be an established infrastructure that supports reevaluation, iteration, and ongoing maintenance. Engaged citizenry is pivotal in building the shared commitment, understanding, and feedback mechanisms necessary for the endurance, adaptability, and responsiveness of social innovations to evolving scales and needs.

However, this study recognises the evident limitation of deriving broad conclusions from a single case. To enhance external validity, particularly regarding the third finding, further research is imperative. The temporal proximity between the conclusion of the Wiener Klimateam project and the drafting of this analysis limits the ability to draw definitive conclusions about the long-term impact of democratic infrastructures on social innovation.

Future research should delve deeper into the role and potential impact of democratic infrastructures on urban social innovation efforts. Despite these limitations, the Wiener Klimateam project and its established democratic infrastructures offer promising evidence of an innovation pace that values and incorporates democratic dialogue and processes.

References

Amatullo M, Boyer B, May J, Shea A (2021) Design for social innovation: case studies from around the world. Routledge

Appio FP, Lima M, Paroutis S (2019) Understanding smart cities: innovation ecosystems, technological advancements, and societal challenges 142:1–14. https://doi.org/10.1016/j.techfore.2018.12.018

Avelino F, Wittmayer JM, Kemp R, Haxeltine A (2017) Game-changers and transformative social innovation. Ecol Soc 22(4). JSTOR. https://doi.org/10.5751/ES-09897-220441

Baste IA, Watson RT, Brauman KI, Samper C, Walzer C (2021) Making peace with nature: a scientific blueprint to tackle the climate, biodiversity and pollution emergencies

Bauler T, Pel B, Backhaus J (2017) Institutionalization processes in transformative social innovation: capture dynamics in the social solidarity economy and basic income initiatives. In: Cohen M,

Szejnwald Brown H, Vergragt P (eds) Social change and the coming of post-consumer society. Routledge, pp 78–94. https://doi.org/10.4324/9781315630168

Concilio G, Li, C, Rausell P, Tosoni I (2019) Cities as enablers of innovation. In: Concilio G, Tosoni I (eds) Innovation capacity and the city: the enabling role of design. Springer International Publishing, pp 43–60. https://doi.org/10.1007/978-3-030-00123-0_3

Dias J, Partidário M (2019). Mind the gap: the potential transformative capacity of social innovation. Sustainability 11(16). https://doi.org/10.3390/su11164465

Ehs T, Zandonella M (2021) Different class citizens: understanding the relationship between socio-economic inequality and voting abstention. Polit Cent Eur 17(3):525–540. https://doi.org/10.2478/pce-2021-0022

Ekins P, Simon S, Deutsch L, Folke C, De Groot R (2003) A framework for the practical application of the concepts of critical natural capital and strong sustainability. Identifying Crit Nat Capital 44(2):165–185. https://doi.org/10.1016/S0921-8009(02)00272-0

European Commission (2019) The European Green Deal. COM/2019/640 final. Communication from the Commission to the European Parliament, the European Council, the Council, the European Economic and Social Committee and the Committee of the Regions. https://eur-lex.europa.eu/legal-content/EN/TXT/?uri=CELEX:52019DC0640

Gordon E, Mugar G (2020) Meaningful inefficiencies: civic design in an age of digital expediency. Oxford University Press

Gordon E, Walter S (2019) Meaningful inefficiencies: resisting the logic of technological efficiency in the design of civic systems. In: Glas R, Lammes S, de Lange M, Raessens J, de Vries I (eds) The playful citizen. Civic engagement in a mediatized culture, vol 1. Amsterdam University Press, pp 310–334. https://library.oapen.org/bitstream/handle/20.500.12657/25946/1004135.pdf?sequence=1#page=311

Grimm R, Fox C, Baines S, Albertson K (2013) Social innovation, an answer to contemporary societal challenges? Locating the concept in theory and practice. Innov: Eur J Soc Sci Res 26(4):436–455. https://doi.org/10.1080/13511610.2013.848163

Hartmann S, Mainka A, Stock WG (2019) Opportunities and challenges for civic engagement: a global investigation of innovation competitions. In: Civic engagement and politics: concepts, methodologies, tools, and applications, pp 607–623. https://doi.org/10.4018/978-1-5225-7669-3.ch030

Haskell L, Bonnedahl KJ, Stål HI (2021) Social innovation related to ecological crises: a systematic literature review and a research agenda for strong sustainability. J Clean Prod 325:129316. https://doi.org/10.1016/j.jclepro.2021.129316

Heikkurinen P, Rinkinen J, Järvensivu T, Wilén K, Ruuska T (2016) Organising in the Anthropocene: an ontological outline for ecocentric theorising. J Clean Prod 113:705–714. https://doi.org/10.1016/j.jclepro.2015.12.016

Hoppe T, De Vries G (2019) Social innovation and the energy transition. Sustainability 11(1). https://doi.org/10.3390/su11010141

Howaldt J, Kaletka C, Schröder A, Zirngiebl M (2019) Atlas of Social Innovation. New practices for a better future

Hubert A (2010) Empowering people, driving change: social innovation in the European Union. European Commission, Bureau of European Policy Advisers, 12 p. https://migrant-integration.ec.europa.eu/library-document/empowering-people-driving-change-social-innovation-european-union_en

Irani L (2019) Chasing innovation: making entrepreneurial citizens in modern India. Princeton University Press

Kılkış Ş (2022) Urban emissions and land use efficiency scenarios towards effective climate mitigation in urban systems. Renew Sustain Energy Rev 167:112733. https://doi.org/10.1016/j.rser.2022.112733

Lee H, Calvin K, Dasgupta D, Krinner G, Mukherji A, Thorne P, Trisos C, Romero J, Aldunce P, Ruane AC (2024) CLIMATE CHANGE 2023 synthesis report: summary for policymakers

Lwasa S, Seto KT (2023) Urban systems and other settlements. In: Intergovernmental Panel on Climate Change (IPCC) (ed) Climate change 2022—mitigation of climate change: working group III contribution to the sixth assessment report of the intergovernmental panel on climate change. Cambridge University Press, Cambridge Core, pp 861–952. https://doi.org/10.1017/9781009157926.010

Mehmood A, Marsden T, Taherzadeh A, Axinte LF, Rebelo C (2020) Transformative roles of people and places: learning, experiencing, and regenerative action through social innovation. Sustain Sci 15(2):455–466. https://doi.org/10.1007/s11625-019-00740-6

Nagesh P, de boer HJ, Dekker SC, van Vuuren DP (2021) Development of scenarios for future emissions of chemicals from agricultural, industrial and urban systems. In: EGU General Assembly conference abstracts, EGU21-1672. https://doi.org/10.5194/egusphere-egu21-1672

Pel B, Haxeltine A, Avelino F, Dumitru A, Kemp R, Bauler T, Kunze I, Dorland J, Wittmayer J, Jørgensen MS (2020) Towards a theory of transformative social innovation: a relational framework and 12 propositions. Res Policy 49(8):104080. https://doi.org/10.1016/j.respol.2020.104080

Reynolds S, Gabriel M, Heales C (2017) Social innovation policy in Europe: where next? (D5.3; SIC Deliverables). https://media.nesta.org.uk/documents/social_innovation_policy_in_europe_-_where_next.pdf

Shabb K, McCormick K (2023) Achieving 100 climate neutral cities in Europe: investigating climate city contracts in Sweden. Npj Clim Action 2(1):6. https://doi.org/10.1038/s44168-023-00035-8

Stake RE (1978) The case study method in social inquiry. Educ Res 7(2):5–8. https://doi.org/10.3102/0013189X007002005

Stake RE (1995) The art of case study research. Sage

Stål HI, Bonnedahl K (2016) Conceptualizing strong sustainable entrepreneurship. Small Enterp Res 23(1):73–84. https://doi.org/10.1080/13215906.2016.1188718

Wittmayer JM, de Geus T, Pel B, Avelino F, Hielscher S, Hoppe T, Mühlemeier S, Stasik A, Oxenaar S, Rogge KS, Visser V, Marín-González E, Ooms M, Buitelaar S, Foulds C, Petrick K, Klarwein S, Krupnik S, de Vries G, Härtwig A (2020) Beyond instrumentalism: broadening the understanding of social innovation in socio-technical energy systems. Energy Res Soc Sci 70:101689. https://doi.org/10.1016/j.erss.2020.101689

Yin RK (1984) Case study research: design and methods. Sage

Social Innovation Design Pathways and Methods Toward Net Zero

Tamami Komatsu◉ and Morgan Ricard◉

Abstract Design pathways allow cities to successfully implement social innovation (SI) within a system of stakeholders committed to achieving climate neutrality. This chapter describes tools and methods that cities and scholars can utilize to develop and enhance social innovation projects. Such methods are based on the Social Innovation Pathway and Toolkit developed in the NetZeroCities project and derived from a design-based learning framework. The approach prioritises human-centred principles of co-design and co-production, and reduces risk by building buy-in, flexibility, nonlinearity, and innovation. Two user pathways are outlined: tailoring the SI process for *bottom-up* and *top-down* support for city Transition Teams, public administration, and local stakeholder working on a city's climate agenda. The lifecycle of the SI pathway is divided into five progressive phases to *analyse the context* for understanding of the system, *reframe problems* while co-designing a portfolio of tools and methods that meet the cities' challenge(s), *envision alternatives* as they take action with iterative ideation, *prototype and experiment* by testing prototypes to gain insight on their impact before full implementation, and *evaluate and scale* to integrate the innovation within the larger ecosystem

Keywords Social innovation · Design pathway · Design toolkit · Learning framework · Co-design · Climate transition

1 Introduction

Social Innovation (SI) is a key lever for cities in achieving climate neutrality, mostly realised through top-down programming and bottom-up action at the urban level (Deserti and Rizzo 2014; Rizzo et al. 2017). Design towards sustainability requires diffused and expert stakeholder efforts (Sevaldson 2011; Manzini 2015) which combine existing innovation elements cut across organisational and disciplinary boundaries, and compel new social relationships (Mulgan et al. 2007). When

T. Komatsu (✉) · M. Ricard
Department of Design, Politecnico di Milano, Milan, Italy
e-mail: tamami.komatsu@polimi.it

© The Author(s) 2025 91
S. Bresciani (ed.), *Social Innovation Projects for Climate Neutral Cities*,
PoliMI SpringerBriefs, https://doi.org/10.1007/978-3-031-87726-1_7

working with public administrators in the urban context, it is necessary for transition teams (i.e., cross-sectoral actors who align local stakeholders) to harness the power of local SI networks and innovators towards their programming efforts (of designing innovative solutions to contextualized challenges within cities' climate action plans and agendas). Impact is dependent on transition teams taking an active role in facilitating knowledge sharing between the city and these local stakeholders.

Within the context of the NetZeroCities (NZC) project, under the EU Mission Cities, designing a SI pathway for cities to follow is a relevant strategy to help cities connect the value of SI along their climate transition journey. NZC's Social Innovation Pathway is based on an iterative, learning-based design process guiding two user pathways (cities' Transition Teams and social innovators) towards implementation of SI initiatives. A Transition Team has been defined by NZC within the Transition Team Playbook (https://netzerocities.app/TransitionPlaybook) (EC 2022) as "*a team spanning across traditional organizational and sectoral boundaries, to create a favourable context for collective action and alignment between local actors.*" Social innovators, on the other hand, are individuals that work to implement new ideas that meet social needs, create social relationships and form new collaborations (EU Commission 2023). These citizens are often quite close to the particular social need, either through direct experience or proximity to affected parties. They can work 'alone' to bring the idea to life but more often work together with a group of individuals and a network of stakeholders. The two user pathways were designed to provide tailored support to (1) the "on-the-ground" and bottom-up initiated SIs led by social innovators responding to emerging needs and (2) to cities' Transition Teams engaged in supporting SI development and amplifying impact through SI Programming. By the latter, we mean that it aims to support city practitioners in amplifying and scaling SI impact, that is, in supporting innovators—within the public administration and all local stakeholders—in bringing their ideas to life through SI. This design provides a holistic framework that facilitates both a top-down perspective of ecosystem setting and impact amplification (transition team) and a bottom-up perspective of solution-building (social innovators).

To accompany the pathway, a toolkit of methods and tools—the Social Innovation Toolkit (https://netzerocities.app/resource-3121)—was designed to navigate users through the different phases of the Social innovation design pathway, equipping them with the necessary tools and methods to achieve the specific objectives of each stage. It was developed by carefully assigning tools and methods that bolster users in implementing each phase of the SI pathway and includes canvases for those tools that required it and lists other tools and methods available on the NZC portal for each phase. The Toolkit was created to help cities design solutions that are inclusive of (1) everyone's needs, both current and future; (2) the lived experience of each system actor; and (3) voices from the margins as an essential means towards designing for all. This is also accomplished by designing for the constraints, by: (1) recognizing the difficulties of changing ingrained social practices; (2) reaching the hard-to-reach; and (3) accounting for system barriers. The toolkit was designed for cities committed to accomplishing the daunting task of achieving climate neutrality by 2030 and for the social innovators seeking to help them in their task.

2 Pathways for Social Innovation Development

Support pathways for SI development can be seen in academic literature from recent years. The Open Book of Innovation (Murray et al. 2010) describes innovation as an ecosystem in which a variety of actors (including policy makers, foundations, philanthropists, social organizations, entrepreneurs, designers, and others) work in tandem toward a common goal, instead of outdated innovation approaches with a single actor working alone. It examines tools and methods of innovation through case studies and theoretical discussions, within a context of intractable social problems, rising costs, and an emerging social economy where existing structures and institutions are transforming to meet new needs. Innovation is elevated from being an effort within the private sector to being the result of combined public and private efforts. The authors describe the future of SI as a constantly evolving and adapting process for which we all share ownership (Fig. 1).

While Murray, Caulier-Grice, and Mulgan explore the larger landscape of SI, other researchers investigate the topic within a singular context. With insight on specific vulnerable communities, Neumeier (2012) argues that SI in rural development contexts often proves superficial and lacking of robust conceptual frameworks.

Neumeier's (2012) Model of Social Innovation (Fig. 2) centres around iteration, multi-actor involvement, contextual sensitivity, with a focus on impact. It makes an argument for continuous learning, adaptation, and improvement, involving multilevel stakeholders along different stages of the process; all while measuring and assessing short- and long-term impacts within their given contexts in a recurring and systematic way. His proposed comprehensive framework offers an actor-oriented network approach to studying SI effectively to better understand the field.

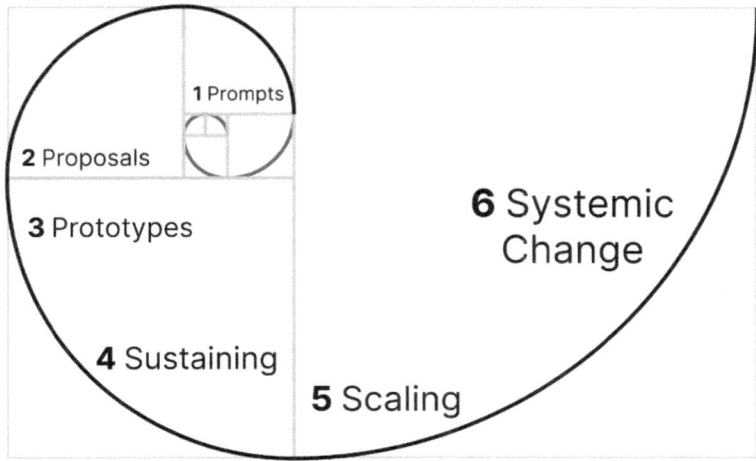

Fig. 1 Social innovation spiral (adapted from Murray et al. 2010)

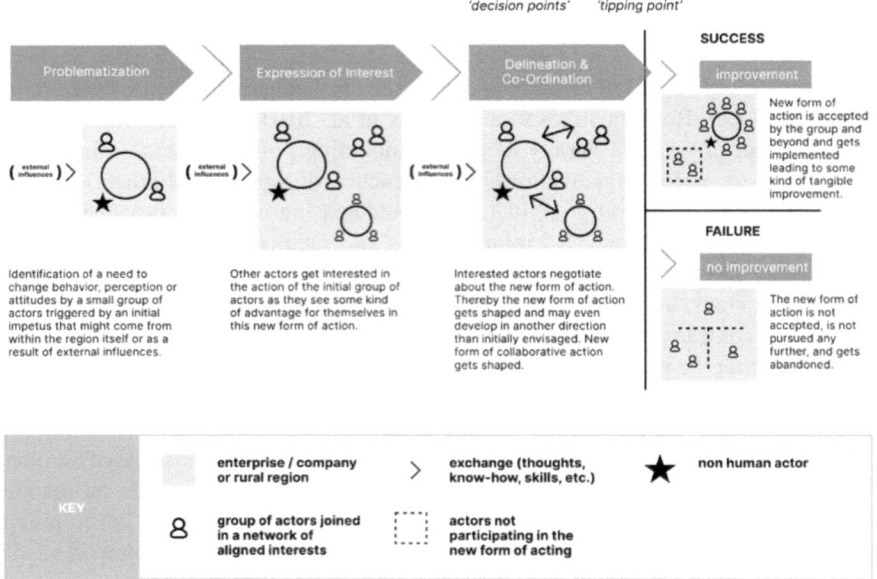

Fig. 2 Model of social innovation (adapted from Neumeier 2012)

When considering collaborative approaches to design processes, Brown and Wyatt (2010) suggest design thinking as an effective problem-solving approach, especially when applied to social issues and comprehension of needs. Starting by *taking inspiration,* the challenge can be understood empathetically by the designers, who attempt to mentally immerse themselves into the perspective of those affected by the issue. *Ideation* then generates a variety of ideas sprung from the insights of the inspiration phase. Creativity is encouraged during brainstorming sessions, without immediate judgement or culling of ideas—the more the merrier until beginning development of the strongest concepts (Kelley and Kelley 2013) through the creation of prototypes for further refinement. The surviving concepts enter *implementation* when prototypes are tested and built upon based on real user feedback and scaled once validated. This requires strategic planning, partnerships, and contextual adaptations.

To implement design thinking on an organisational level of systems and cultures, Elsbach and Stigliani (2018) identify emotional experiences and physical artefacts as part of a sensemaking strategy for cultural change. Their highly cited review of the empirical research on using design thinking within organisations highlights experiential learning processes (Kolb 2014) as essential to learning.

Their framework (Fig. 3) has been applied to organisations to enhance creativity, improve collaboration, and fuel continuous learning cycles. Within projects for climate neutral cities, administrators must integrate learning experiences into their existing operations and culture for successful implementation of culture-shifting innovations.

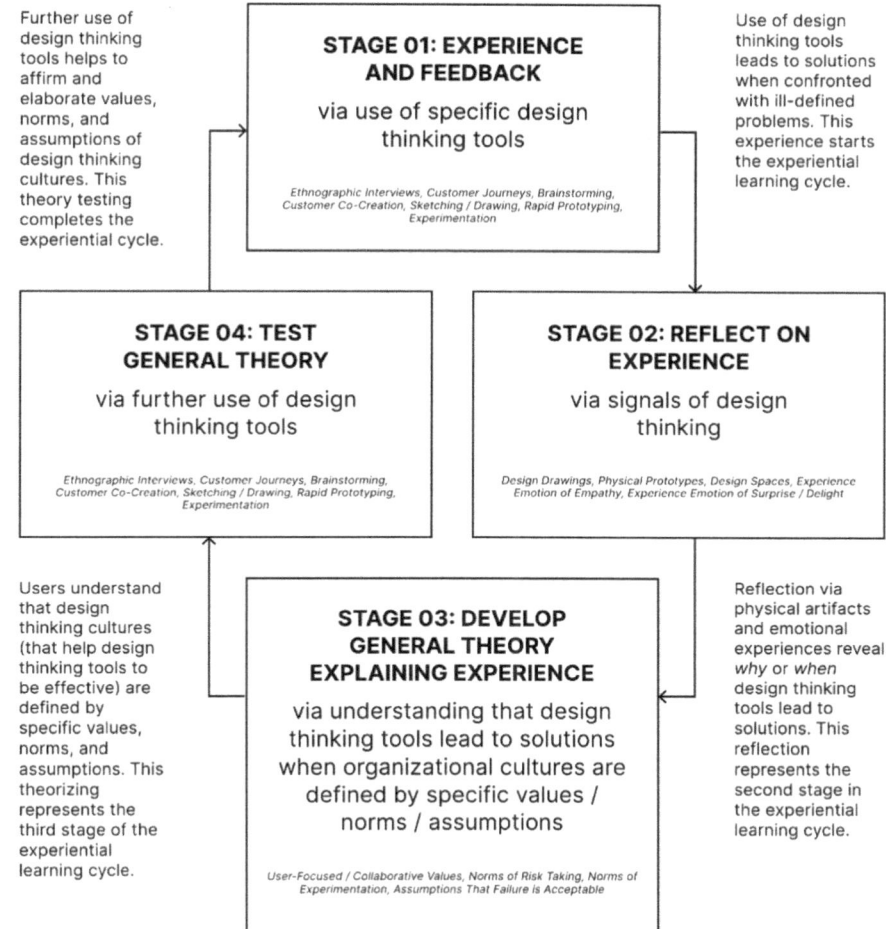

Fig. 3 Experiential learning framework relating design thinking tools and cultures (adapted from Elsbach and Stigliani 2018)

A design-based learning framework (Rizzo et al. 2017) was selected to establish human-centred principles of co-design and co-production, incorporating learning goals and learning through experimentation. The framework was iteratively adapted and designed to de-risk innovation, build buy-in, and foster an open-ended innovation process. It offers context-sensitive flexibility for different user types, instilling confidence in its applicability and an emphasis on non-linear iteration. Users are encouraged to embrace the dynamic nature of the research and learning process for top-down and bottom-up solutions across the system of actors.

The Social Innovation Pathway, developmentally following Kolb's (1984) model of experiential learning, is compatible with the NZC Climate Transition Map (CTM, described in Chap. 3). Kolb's model was chosen for two main reasons: (1) learning by doing is an innate part of the design process, often making design an implicit agent of change, and (2) the urgency to act, which is common to most 'wicked' challenges, often requires fast action and to learn while trying to solve the problem quickly and iteratively. By adopting an experiential learning approach—"the process whereby knowledge is created through the transformation of experience" (Kolb 2014, p. 41)—both social innovators and cities are encouraged to learn quickly through experience and reflect while acting (Schön 1983). Kolb's approach, similar to the design process described further below, passes through four stages, alternating between two realms: that of theory and that of experience, as in other models (Owen 1998; Argyris and Schön 1978, 1996). The four phases of his model are as follows: concrete experience, reflective observation, abstract conceptualisation, and active experimentation. As visualized in Fig. 4, the phases of the SI pathway can be found in the space between Kolb's learning cycle. While not a linear process, the pathway can be seen to "start" by *analyzing the context*, then *reframing the problem, envisioning alternatives* and then *developing prototypes* and iterating. *Evaluate and Scale* is an underpinning activity of the entire pathway, but specifically takes social innovators and cities from the concrete experience of testing solutions through prototypes to envisioning new alternatives based on the feedback and insights gained from experimentation. In designing the framework, the objective is to encourage cities and social innovators to act and build knowledge through experience while responding to their need for urgency.

The SI pathway (Fig. 4) captures the SI lifecycle in these ways: analyse context, reframe problems, envision alternatives, prototype and experiment and evaluate and scale, loosely following the double diamond design process (Design Council 2005).

3 SI Implementation Process Flow

1. Implementing a SI initiative starts with a divergent discovery phase that explores the challenge in context, from the needs of the local communities, to the stakeholders that could be involved and the resources available.
2. This is followed by a convergent phase that (re-)defines the challenge according to the insights found in the discovery phase. After this first stage of divergence and convergence in defining the existing challenge (the first "diamond" in the double diamond process), comes a second round to define the solution (the second "diamond"). Active participation in this phase is crucial, as it empowers you to shape the direction of the initiative.
3. The next divergent phase envisions various solutions for the challenge—services, products, organizational models, network formations, etc.
4. It is followed by a second convergent phase that selects ideas or prototypes and tests them for validity, feasibility and impact.

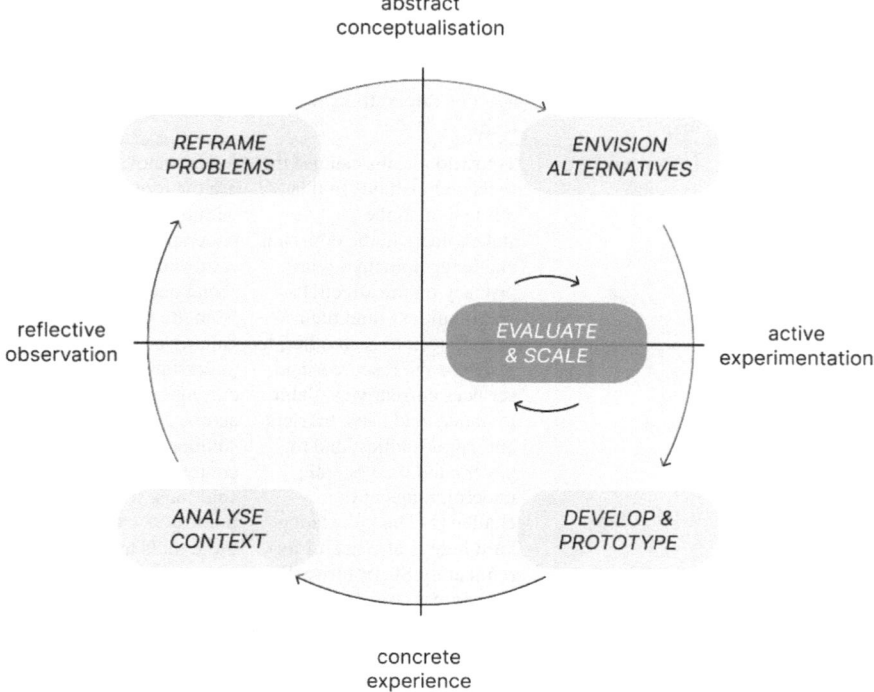

Fig. 4 NZC's social innovation pathway, adapted from Rizzo et al. (2017)

5. Finally, the last phase is dedicated to evaluating and scaling the solutions. While the entire process is iterative and non-linear, this last phase is a continuous activity that creates synthesis but also strategic insight for future development of the initiative.

To further understand each phase of the Social Innovation User Pathways (Table 1), we outline each in terms of its purpose, objective, addressing of cities' concerns, and finally a list of tools and methods which can be accessed through an interactive graph (Fig. 5; available at https://netzerocities.app/resource-4324).

3.1 Analyse the Context

It is an essential step towards building solutions and strengthening ecosystems. It is about understanding the context, both in 'hard' terms—the infrastructure of people, organizations, companies, spaces, norms and regulations—and 'soft' terms—i.e. the practices, routines and beliefs that inform everyday life and the choices we make. This phase explores these contextual factors, their inter-relationship and how

Table 1 NZC's social innovation pathway user pathways matched to the climate transition map described in Chap. 3 strategizing for climate neutrality

Phase	Relation to the climate transition map	Use for (public administrators) transition team	Use for (citizen) social innovators
1. Analyse the context	Understand the system	Transition teams can use the tools and methods in this phase to map the stakeholders in the emission challenge domain – with primacy on the affected communities – and their connections to each other; to visualise the resources and services currently available to understand gaps, barriers and opportunities; and to understand the systemic underpinnings of the challenge. The knowledge built here is also useful for replicating SIs or for scaling their impact by identifying the contextual elements that led to their success	Social innovator(s) can use the tools and methods within the map to understand better the context of the specific social need emerging from the transition by gaining a systemic understanding of the current ecosystems of actors, resources and solutions. As context-dependent solutions, social innovators can also use these tools to understand how to replicate solutions found elsewhere in their own city or neighbourhood
2. Reframe problems	Understand the system; co-design a portfolio	Transition teams can use the tools and methods in this phase to define more tailored challenge questions that reflect the needs of affected communities of stakeholders; identify areas where social innovators can be engaged and/or innovation areas whose impact should be amplified; and how to create future-fit strategies that are inclusive and effectively respond to real needs (current and emergent)	Social innovator(s) can use the tools and methods to further refine their challenge statement and solution to provide more effective services or to update their current services to better reflect current and new needs; to understand new user segments; align value propositions to be future fit; and to amplify their current offer to diversify their impact

(continued)

Table 1 (continued)

Phase	Relation to the climate transition map	Use for (public administrators) transition team	Use for (citizen) social innovators
3. Envision alternatives	Co-design a portfolio, take action	Transition teams can use the tools and methods in this phase to create new solutions to include in their portfolio of actions that serve new user segments; find new mechanisms to align multi-stakeholder interests; ideate policy actions/programs that is inclusive and enabling; define new governance models to achieve the mission; and find new combinations to empower more effective collaboration and action	Social innovator(s) can use the tools and methods to ideate new solutions to fill impact gaps or meet emerging needs; discover where powerful alliances can be made; and improve service offer to align with the mission's value proposition
4. Prototype and experiment	Take action; learn and reflect	Transition teams can use the tools and methods in this phase to test new solutions, organisational models, and/or network formations by implementing certain features or testing specific 'touchpoints' of the solution; to gain insight on the effectiveness and impact of solutions before full implementation; to set up emission domain ecosystems; and to engage in double-loop learning unlocking potentially opportunities for transformational change	Social innovator(s) can use the tools and methods to test solutions for validity, impact and feasibility; to gain buy-in and commitment from stakeholders and potential users; to test different pathways of development; and test for impact
5. Evaluate and scale	Make it the new normal	Transition teams can use the tools and methods in this phase to evaluate social innovation initiatives and explore ways to support them to scale up	Social innovator(s) can use the tools and methods to evaluate themselves and find suitable strategies or methods to scale up their initiatives

How to make social innovation relevant in the city's journey to climate neutrality

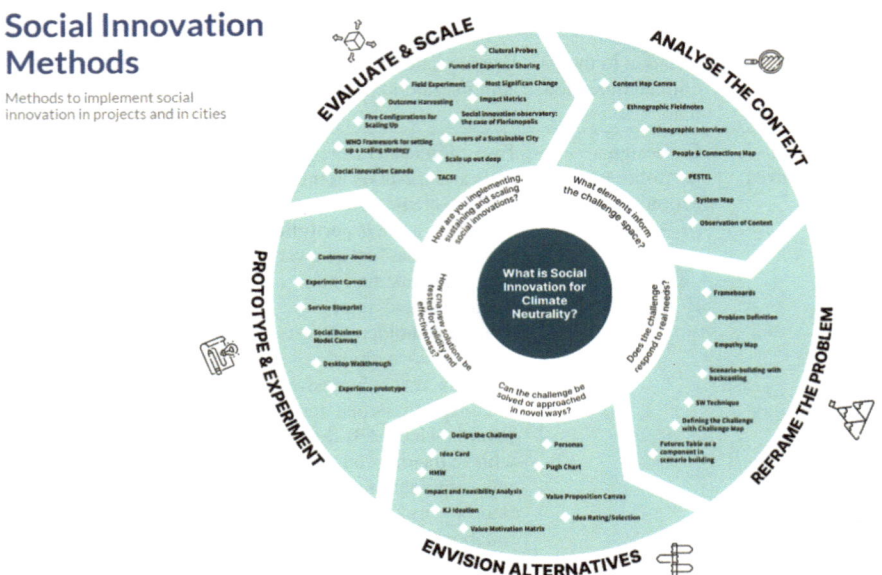

Fig. 5 Interactive graph of social innovation methods mapped by category. *Source* https://netzer ocities.app/resource-4324

they influence the challenge space. SI responds to unmet social needs; this requires understanding the need from multiple perspectives. This stage can help cities respond to the following questions:

- What elements inform the challenge space?
- What is my city already doing in SI for Climate Neutrality (e.g. policies, funding programs, training centres)? How is the need currently being met?
- What are the specific needs of citizens and other actors, particularly the marginalized, in the transition to climate neutrality?
- What resources are available?
- Which actors could be engaged in my climate goals?
- Which actors gravitate around the need?
- What resources are available to those in need or for other service providers?

In their journey to climate neutrality, cities will need to identify priority emission reduction challenges: SI is a lever of change to address these challenges. Cities will need to onboard diverse communities of stakeholders with different needs, priorities and capacities to act. To properly scope the challenge—i.e. the affected communities, the barriers, gaps and impact—cities will need to understand and map the ecosystem of actors and resources that gravitate around these challenges to mobilize them successfully around the mission. This fits into the *Understand the System* phase of the Climate Transition Map and when done well can also act as a strategic step of drafting an effective Climate City Contracts that can truly engage diverse stakeholders and activate multi-actor collaborations by building a shared value proposition upon which different stakeholders and platforms can align.

Complex problems, such as Mission challenges (Mazzucato 2018), are often experienced and understood in different ways by different actors. Translating larger mandates into local contexts and needs requires pooling together different actors to reframe the challenge. The process not only deepens understanding of the challenge but also provides insight into the current system and how it can be improved, generating several insights for innovation on different time horizons. Sometimes, it is helpful to look at the present from the perspective of the future to ensure that what we are doing now will fit into the future we want. These future scenarios and visions also work to include the voices of future generations in the solution-building process.

Tools and Methods used to Analyse the Context:

1. Context Map Canvas (https://netzerocities.app/resource-3035): A visualisation tool for an internal understanding amongst the project teams, and to help understand trends and different perspectives.
2. Ethnographic Fieldnotes (https://netzerocities.app/resource-3040): For organising observations, analysis, questions, reflections, and ideas for future action.
3. Ethnographic Interview (https://netzerocities.app/resource-3041): for deep understanding of actions and motivations through the relationships and connections between the researcher and the community they work within.
4. People and Connections Map (https://netzerocities.app/resource-3059): A visualisation tool for identifying stakeholders and how they will be reached, this tool maps surrounding actors who could be brought within the system.
5. PESTEL (https://netzerocities.app/resource-3061): An analysis tool for assessing the political, economic, social, technological, environmental, and legal forces which face an organisation.
6. Systems Map (https://netzerocities.app/resource-3064): A schematic representational tool showing the main actors of a system from the POV of the primary service-providing organisation, which serves as a living document that will change over time as the system evolves.
7. Empathy Map (https://netzerocities.app/resource-3039): A visualisation tool used to collaboratively externalize knowledge about a particular user in order to build a shared understanding of needs and insights for decision making purposes.
8. Influencing Factors Matrix (https://netzerocities.app/_content/files/knowledge/3121/social_innovation_toolkit_compressed.pdf): A visualisation tool of individual actors and actor groups to understand their motives to become engaged in a solution; particularly the positive aspects, negative aspects, barriers, and obstacles they might face.
9. Personas (https://netzerocities.app/resource-3060): A visualisation tool to define the dimensions and distinguishing characteristics of a customer segment to inform development of a product or service experience.

10. Customer Journey Map (https://netzerocities.app/resource-3037): A visualisation tool from the perspective of the user which describes each step of interaction made within the organization or system.

3.2 Reframe the Problems

It is a phase of convergence, where all knowledge and experience of the challenge are synthesised into insights that inform the creation of a more refined challenge(s). This may come from activities that analyse the context (See *Analyse the Context* phase) or from built-up knowledge and experience of the specific challenge. Social innovators, for example, often skip the stage of need analysis and user research since they possess in-depth knowledge and (personal) experience with the challenge (Rizzo et al. 2017).

The objective of the phase is to redefine the challenge questions to ensure that real and existing needs are being effectively addressed. From the city perspective, policy ecosystems are often built around a single-user mentality based on population averages rather than being crafted on specific user groups or segments. The dilemma between representation and validity in serving populations in need is a common struggle for city administrations. As such, it is an opportunity for cities to engage social innovators and amplify their solutions to deliver effective and more tailored solutions to citizens. Reframing the challenge is a key step in creating more tailored solutions for affected communities of stakeholders. The phase connects the *Understand the System* and *Co-Design Portfolio* phases of the Climate Transition Map, as it transforms knowledge from the context into insights that can inform cities in building portfolios that respond to multiple needs and represent diverse populations. It can help cities respond to the following questions:

1. Does the challenge respond to real needs?
2. How does my city plan to achieve net zero emissions in a systemic, inclusive and anticipatory manner?
3. What is the societal challenge being addressed?
4. How can my city respond to the specific needs while achieving climate goals?
5. How can SI contribute to co-benefits of net zero emission?
6. Are my climate goals future-fit?
7. How can existing SI be useful towards the city's climate goals?

Tools and Methods used to Reframe Problems:

1. Frameboards (https://netzerocities.app/resource-3053): A visualisation and communication tool used to portray a certain temporary perspective on a problem or challenge being explored.
2. Problem Definitions (https://netzerocities.app/resource-3062): The first stage of developing a response to a problem, during which a group establishes shared comprehension of underlying cause(s) of a symptomatic issue.

3. Empathy Map (https://netzerocities.app/resource-3039): A visualisation tool for collaboratively externalize knowledge about a particular user in order to build a shared understanding of needs and insights for decision making purposes.
4. People and Connections Map (https://netzerocities.app/resource-3059): A visualisation tool for identifying stakeholders and how they will be reached, this tool maps surrounding actors who could be brought within the system.
5. Designing the Challenge (https://netzerocities.app/resource-3038): A tool which transforms the challenge into an innovation competition that attracts an audience of solution makers guided by outlined ambitions and constraints, encouraging bounded creativity for fit-for-purpose responses.
6. How Might We (https://netzerocities.app/_content/files/knowledge/3121/soc ial_innovation_toolkit_compressed.pdf): A user-centered problem-solving method that empowers a team to reframe complex problems as open-ended questions.
7. Influencing Factors Matrix (https://netzerocities.app/_content/files/knowledge/ 3121/social_innovation_toolkit_compressed.pdf): A visualisation tool of individual actors and actor groups to understand their motives to become engaged in a solution; particularly the positive aspects, negative aspects, barriers, and obstacles they might face.
8. Motivation Matrix (https://netzerocities.app/resource-3058): An exercise that assists facilitators and designers in measuring the motivation of actors based on six core motivation factors: incentive, achievement, social acceptance, fear, power, and growth.
9. Personas (https://netzerocities.app/resource-3060): A visualisation tool to define the dimensions and distinguishing characteristics of a customer segment to inform development of a product or service experience.
10. Service Blueprint (https://netzerocities.app/resource-2343): A visualisation tool that elevates the business model by providing the big picture of processes by which value is created, delivered, and captured.

3.3 Envision Alternatives

Faced with the challenge of reaching climate neutrality by 2030, cities must find alternative ways of doing things and measures to stimulate multi-actor collaboration across sectors. This means breaking away from silo mentality and working practices, whether that be within the city administration or between sectors or clusters. Co-creating a shared vision for the city can be a useful and strategic action to onboard urban stakeholders. Aligning value propositions to implement the vision is a natural consequence of this commitment and requires creating different pathways of collaboration and development. This can lead to the creation of new solutions or crafting new combinations or formations of existing offers. Equipped with a deep understanding of the context and the challenge, the phase is dedicated to generating new ideas based on previous reflection, dialogue and insights into the challenge.

Envisioning alternatives is not only about ideation in terms of new solutions but also new governance models, economic models, ecosystem arrangements, constellations of actors, etc. Having mapped the emission challenge's ecosystem of actors and current solutions, cities can find new ways to empower existing solutions (e.g., through new policies, creating knowledge-sharing platforms, creating new connections, etc.) to serve new stakeholder categories or deliver existing services more effectively.

As a divergent phase in the design process, the step explores all possible solutions, typically with lesser concern for the idea's feasibility. The phase ends in a moment of synthesis where the most promising ideas, based on their impact and feasibility, are selected to move on to the prototyping phase. Envisioning alternatives connects The *Co-Design a Portfolio* and *Take Action* phases of the Climate Transition Map as it can offer new solutions—services, products, organizational models, governance structures—to integrate with a city's portfolio and can help cities answer the following question:

- Can the challenge be solved or approached in novel ways?
- What new solutions are needed to bring my city on an inclusive and effective path towards net zero emissions?
- How can the city ideate new ways to align interests around decarbonisation goals?
- How can the city design policy frameworks for climate targets that include the specific needs of its citizens and the city's other actors?
- How can the city empower multi-actor collaboration through new governance models or ecosystem support?
- What new combinations of existing solutions could contribute to greater collective impact?
- What network formations could empower more effective collaboration and action?

Tools and Methods used to Envision Alternatives:

1. Designing the Challenge (https://netzerocities.app/resource-3038): A tool which transforms the challenge into an innovation competition that attracts an audience of solution makers guided by outlined ambitions and constraints, encouraging bounded creativity for fit-for-purpose responses.
2. Idea Card (https://netzerocities.app/resource-3056): A single-page organisation tool to detail the challenges, needs, functions, and actors involved in an idea.
3. How Might We (https://netzerocities.app/_content/files/knowledge/3121/soc ial_innovation_toolkit_compressed.pdf): A user-centered problem-solving method that empowers a team to reframe complex problems as open-ended questions.
4. Influencing Factors Matrix (https://netzerocities.app/_content/files/knowledge/3121/social_innovation_toolkit_compressed.pdf): A visualisation tool of individual actors and actor groups to understand their motives to become engaged in a solution; particularly the positive aspects, negative aspects, barriers, and obstacles they might face.

5. Impact-Feasibility Matrix (https://netzerocities.app/resource-3082): A tool for prioritising and deciding upon ideas/projects efforts and timelines.
6. KJ Ideation: A brainstorming method of collecting, sorting, and finding meaning within qualitative data.
7. Motivation Matrix (https://netzerocities.app/resource-3058): An exercise that assists facilitators and designers in measuring the motivation of actors based on six core motivation factors: incentive, achievement, social acceptance, fear, power, and growth.
8. Personas (https://netzerocities.app/resource-3060): A visualisation tool to define the dimensions and distinguishing characteristics of a customer segment to inform development of a product or service experience.
9. Pugh Chart (https://netzerocities.app/resource-3063): A tool for comparison of options according to the ranking of specific needs and values contextualised to a specific situation.
10. Value Proposition Canvas (https://netzerocities.app/resource-3065): A tool to establish the most important aspects of a service or product being built and tested, and what opportunities exist to provide value to users.
11. People and Connections Map (https://netzerocities.app/resource-3059): A visualisation tool for identifying stakeholders and how they will be reached, this tool maps surrounding actors who could be brought within the system.
12. Systems Map (https://netzerocities.app/resource-3064): A schematic representational tool showing the main actors of a system from the POV of the primary service-providing organisation, which serves as a living document that will change over time as the system evolves.
13. Frameboards (https://netzerocities.app/resource-3053): A visualisation and communication tool used to portray a certain temporary perspective on a problem or challenge being explored.
14. Empathy Map (https://netzerocities.app/resource-3039): A visualisation tool for collaboratively externalize knowledge about a particular user in order to build a shared understanding of needs and insights for decision making purposes.

3.4 Prototype and Experiment

As 'wicked' challenges, mission challenges are hard to solve because of the highly interconnected and systemic nature of the problems. Testing solutions to complex challenges can often mean creating system-level prototypes that require high investments of time and capital. The tools in this phase are meant to help prototype certain features or specific interactions happening at different 'touchpoints' of the solution, helping to ensure that the solutions are purposefully built around life experience and concrete needs to provide real value. Prototyping also helps de-risk innovations by not only attempting to work out problems pre-emptively but also by learning by doing and building up the knowledge needed to implement the innovation. After prototyping, agile piloting and experimentation can take the solutions a step further.

On a more macro-scale, SIs can be seen to act as small-scale experiments and prototypes of scaled solutions (Rizzo et al. 2017). In practice, the innovations act as 'boundary objects' for diverse stakeholders to come around (Star and Griesemer 1989; Dorst 2015). This increases buy-in and facilitates implementation. The insights coming from this important phase feed into future re-framing of the challenges and the co-evolution of the solution (Dorst 2019) until the best 'fit' is found between the challenge space and the solution space. This is an important phase of the process that pushes for iteration and double-loop learning (Argyris and Schön 1978, 1996), asking the city administration or organization to question the assumptions, norms, principles and values that underpin its current operation.

The phase connects the *Take Action* and *Learn and Reflect* phases of the Climate Transition Map, as it asks cities and social innovators to actively implement parts or scales of the solution to test for validity, impact and feasibility and to learn from the insights to either refine the solution or to question the original frame of the challenge space. The stage helps cities answer the following questions:

1. How can new solutions be tested for validity and effectiveness?
2. How can the city test SI before scaling and making large infrastructural changes?
3. How can specific features be more effective and people-centred?
4. Does the service/product really satisfy the needs of the target user?
5. How can the city experiment with SI ideas?
6. What assumptions, norms, practices and/or values are put into question as a result of the prototype?

Tools and Methods used to Prototype & Experiment

1. Customer Journey Map (https://netzerocities.app/resource-3037): A visualisation tool from the perspective of the user which describes each step of interaction made within the organization or system.
2. Experiment Canvas (https://netzerocities.app/resource-3042): A tool for creating experiments to test ideas for a specific topic.
3. Service Blueprint (https://netzerocities.app/resource-2343): A visualisation tool that elevates the business model by providing the big picture of processes by which value is created, delivered, and captured.
4. Social Business Model Canvas (https://netzerocities.app/resource-2632): A visualisation tool to advance a business model to ensure the processes create, deliver, and capture value.
5. Ethnographic Fieldnotes (https://netzerocities.app/resource-3040): For organising observations, analysis, questions, reflections, and ideas for future action.
6. Ethnographic Interview (https://netzerocities.app/resource-3041): for deep understanding of actions and motivations through the relationships and connections between the researcher and the community they work within.
7. Systems Map (https://netzerocities.app/resource-3064): A schematic representational tool showing the main actors of a system from the POV of the primary service-providing organisation, which serves as a living document that will change over time as the system evolves.

8. Frameboards (https://netzerocities.app/resource-3053): A visualisation and communication tool used to portray a certain temporary perspective on a problem or challenge being explored.

9. Motivation Matrix (https://netzerocities.app/resource-3058): An exercise that assists facilitators and designers in measuring the motivation of actors based on six core motivation factors: incentive, achievement, social acceptance, fear, power, and growth.

10. Pugh Chart (https://netzerocities.app/resource-3063): A tool for comparison of options according to the ranking of specific needs and values contextualised to a specific situation.

11. Value Proposition Canvas (https://netzerocities.app/resource-3065): A tool to establish the most important aspects of a service or product being built and tested, and what opportunities exist to provide value to users.

12. Funnel of Experience Sharing (https://netzerocities.app/resource-908): A visual template to facilitate learning extraction from experiences.

3.5 Evaluate and Scale

While evaluation is often thought of as a post-implementation activity, it is useful to know how to evaluate solutions from the beginning to design truly impactful solutions. Measuring impact becomes a strategic asset for understanding effectiveness and knowing what, when and how to adapt the solution for a better fit or to scale the solution for wider impact.

This phase answers the following questions:

- How are you implementing, sustaining and scaling SI?
- How can the city evaluate current SI initiatives as prototypes to be scaled?
- How can SI be scaled up?
- How can SI be evaluated?
- Does the SI fit all the user criteria?
- What solutions already exist that could be scaled or empowered through policy?

Tools and Methods used to Evaluate & Scale:

1. Funnel of Experience Sharing (https://netzerocities.app/resource-908): A visual template to facilitate learning extraction from experiences.

2. Ethnographic Interview (https://netzerocities.app/resource-3041): for deep understanding of actions and motivations through the relationships and connections between the researcher and the community they work within.

3. Influencing Factors Matrix (https://netzerocities.app/_content/files/knowledge/3121/social_innovation_toolkit_compressed.pdf): A visualisation tool of individual actors and actor groups to understand their motives to become engaged in a solution; particularly the positive aspects, negative aspects, barriers, and obstacles they might face.

4. Impact-Feasibility Matrix (https://netzerocities.app/resource-3082): A tool for prioritising and deciding upon ideas/projects efforts and timelines.
5. Pugh Chart (https://netzerocities.app/resource-3063): A tool for comparison of options according to the ranking of specific needs and values contextualised to a specific situation.
6. Experiment Canvas (https://netzerocities.app/resource-3042): A tool for creating experiments to test ideas for a specific topic.
7. Service Blueprint (https://netzerocities.app/resource-2343): A visualisation tool that elevates the business model by providing the big picture of processes by which value is created, delivered, and captured.
8. Social Business Model Canvas (https://netzerocities.app/resource-2632): A visualisation tool to advance a business model to ensure the processes create, deliver, and capture value.

To surmise, as an iterative process, cities and social innovators may enter at any phase and can always be redirected to go back to previous stages based on emerging insights.

4 Conclusion

The SI Pathway and toolkit have been actively used by NZC in its seasonal schools. NZC Seasonal Schools aim to support cities meet their operational and strategic needs through capacity building. Each edition of the school seeks to meet the cities' current needs, which are changing as they move forward on their transition journey. The SI pathway and tools have therefore been used thus far mainly to support strategy design via the climate city contracts and action plans. For example, tools like the Problem Definition and People and Connections Map have been used to help cities analyse their context and understand the resources available, while the idea card has been used to help ideate possible solutions to add, for instance, to their action plans. As the Cities Mission progresses, further research should focus on how the tools and methods will be used in the implementation of the cities' action plans.

References

Argyris C, Schön DA (1978) Organisational learning: a theory in action perspective. Addisson Wesley, USA
Argyris C, Schön D (1996) Organizational learning II: theory, method and practice. Addison-Wesley
Brown T, Wyatt J (2010) Design thinking for social innovation. Dev Outreach 12(1):29–43
Deserti A, Rizzo F (2014) Design and organizational change in the public sector. Des Manag J 9(1):85–97
Design Council (2005) A study of the design process—the double diamond
Dorst K (2015) Frame creation and design in the expanded field. She Ji: J Des Econ Innov 1(1):22–33

Dorst K (2019) Co-evolution and emergence in design. Des Stud 65:60–77. https://doi.org/10.1016/j.destud.2019.10.005

Elsbach KD, Stigliani I (2018) Design thinking and organizational culture: a review and framework for future research. J Manag 44(6):2274–2306

EU Mission Cities (2022) Transition team playbook https://netzerocities.app/TransitionPlaybook

European Commission (2023) Social innovation. In: Internal market, industry, entrepreneurship and SMEs

Kelley T, Kelley D (2013) Creative confidence: unleashing the creative potential within us all, 1st edn. Crown Business, New York

Kolb DA (1984) Experiential learning: experience as the source of learning and development. Prentice-Hall, Inc

Kolb DA (2014) Experiential learning: experience as the source of learning and development. FT Press

Manzini E (2015) Design, when everybody designs: an introduction to design for social innovation. MIT Press

Mazzucato M (2018) Mission-oriented innovation policies: challenges and opportunities. Ind Corp Chang 27(5):803–815

Mulgan G, Tucker S, Ali R, Sanders B (2007) Social innovation: what it is, why it matters, how it can be accelerated. University of Oxford, Young Foundation, London. Retrieved December 14, 2023 from https://youngfoundation.org/wp-content/uploads/2012/10/Social-Innovation-what-it-is-why-it-matters-how-it-can-beaccelerated-March-2007.pdf

Murray R, Caulier-Grice J, Mulgan G (2010) The open book of social innovation, vol 24. Nesta, London

Neumeier S (2012) Why do social innovations in rural development matter and should they be considered more seriously in rural development research? –proposal for a stronger focus on social innovations in rural development research. Sociol Rural 52(1):48–69

Owen C (1998) Design research: building the knowledge base. Des Stud 19:9–20

Rizzo F, Deserti A, Cobanli O (2017) Introducing design thinking in social innovation and in public sector: a design-based learning framework. Eur Public Soc Innov Rev 2(1):127–143

Schön D (1983) The reflective practitioner: how professionals think in action. Basic Books

Sevaldson B (2011) GIGA-mapping: visualisation for complexity and systems thinking in design. Nordes 4:49

Star SL, Griesemer JR (1989) Institutional ecology, 'translations' and boundary objects: amateurs and professionals in Berkeley's museum of vertebrate zoology, 1907–39. Soc Stud Sci 19(3):387–420

Advancing Social Innovation for a Carbon–Neutral Future: A Portfolio-Based Approach for the New European Bauhaus

Emma Puerari⊙ and **Alessandro Deserti**⊙

Abstract This chapter explores the New European Bauhaus (NEB) EU research programme as a relevant instrument for advancing the European Green Deal by fostering tangible and on-the-ground improvements in people's quality of life, mixing social innovation with more traditional innovation fields and approaches. The NEB programme, launched by the European Commission in 2020, brings forward a set of initiatives by integrating sustainability, inclusivity, and aesthetics. By leveraging partnerships across public, private, and civil society sectors, the NEB promotes small-scale, adaptable initiatives that collectively contribute to climate neutrality. The urgency of building a carbon–neutral future is framed as a complex problem that addresses technological, environmental, and social responsibility, given its impact on different sectors such as healthcare, education, mobility, housing, and overall well-being. Social innovations are positioned as essential to tackle the challenges bound to these interconnected sectors. However, exploiting social innovation to achieve systemic change requires dismantling existing barriers between sectors, governance levels, and cultural structures, and reframing the traditional "scaling" mechanisms in a broader perspective that combines bottom-up and top-down approaches. The chapter outlines how the NEB programme relates to social innovation for climate neutrality. Secondly, it emphasises the importance of adopting mission-oriented and portfolio-based approaches in EU-funded research schemes to ensure sustainable transitions. The chapter concludes with an analysis of the progress made by the NEB and the challenges that remain to include social innovation among the fundamental means to meet the grand societal goals of the European Green Deal and EU Missions.

Keywords EU Green Deal · EU Missions · Impact monitoring assessment and learning · Scaling mechanisms · NEB values and principles

E. Puerari (✉) · A. Deserti
Department of Design, Politecnico di Milano, Milan, Italy
e-mail: emma.puerari@polimi.it

A. Deserti
e-mail: alessandro.deserti@polimi.it

S. Bresciani (ed.), *Social Innovation Projects for Climate Neutral Cities*,
PoliMI SpringerBriefs, https://doi.org/10.1007/978-3-031-87726-1_8

1   Introduction

The urgent need to build a carbon–neutral future is crucial, as it directly impacts the quality of life of urban populations, impacting a variety of interconnected factors, such as health, education, mobility, housing, employment and overall well-being. The pursuit of sustainability is not only an environmental objective but also a social responsibility. The climate crisis disproportionately affects vulnerable populations, worsening inequalities and threatening access to basic resources like food, clean water, and healthcare. Since these aspects are linked to broader societal systems, improving them requires collective action rather than individual efforts. Therefore, tackling the complexities of the quality of life necessitates more than fragmented solutions; it calls for implementing social innovation initiatives that can bring about systemic transformation.

Social innovations, as defined in Chap. 1, are novel social practices, that comprise new ideas, products, services, governance models, rules and social relations. According to Franz et al. (2012), social innovations stand distinct from other forms of innovation as they necessitate not only operational or technological support, but also require profound shifts in individual and organisational behaviours, norms, culture, beliefs and values. Over recent decades, numerous activists, networks, private companies, research institutes, NGOs and local governments worldwide have been actively involved into diverse social innovation initiatives tailored to specific contexts, often operating at the local scale. These initiatives confront persistent issues within current economic systems while striving to establish concrete alternative solutions that could impact existing development structures. Although recognised as crucial for climate neutrality, the process to scale up, out and deep (Moore et al. 2015) social innovation initiatives is accompanied by several difficulties as described in Chap. 5. The very idea that social innovation can create impact only by leveraging scaling mechanisms should be reconsidered, starting from a better understanding of the social innovation process based on the analysis of real-life cases (Rizzo et al. 2020; Healy et al. 2024). To have impact on grand societal challenges social innovation projects do not only require activism from specific networks, partnership development, institutional and governmental support, but also to dismantling existing barriers between sectors, departments, governance levels, power dynamics and inner cultural structures (Hölscher et al. 2018). Addressing such complexity requires a deep understanding of how socio-technical and socio-technological systems function and co-evolve towards unknown statuses (De Roo 2018) and the capacity to navigate transitions towards sustainability and climate neutrality. The adoption of mission-oriented (Mazzucato 2018a; b) and systemic approaches has been highlighted as crucial to address transitions towards sustainability (Geels 2002). Such approaches share the need for audacious long-term goals and for a portfolio-based approach (Mazzucato et al. 2024) that spans across different levels of governance and practice. The scale of the problems to be tackled calls for systematic initiatives, and the portfolio-based approach of funding schemes is emerging as a potential response since it involves a diverse set of coordinated actions across sectors, where public, private, and civil society actors work together to foster innovation and experimentation (UNDP 2022). Instead of focusing on a large-scale approach with fixed actions to achieve outcomes, portfolios advocate for a complementary array of smaller, adaptable initiatives allowing flexibility in response to complex challenges. Aligned with this strategy, Mazzucato (2018a; b) emphasises the importance of adopting "mission-oriented" policies, where governments set clear goals (missions), while fostering a diverse range of experiments and innovations to flourish under that mission's umbrella. The purpose behind adopting such an approach is to allow for the necessary flexibility for a successful portfolio to combine risk-taking and long-term vision, experimentation and stability, learning from failures as much as from successes, aligning decarbonization and economic growth with social and environmental outcomes. The concept, coming directly from the recognition of the difficulty of measuring the impacts of small-scale initiatives such as social innovation projects, is particularly relevant when related to grand societal challenges such as building a carbon–neutral future, where a variety of solutions must co-evolve towards common results.

The New European Bauhaus (NEB) EU research funding programme builds on these foundations to advance the implementation of the European Green Deal on the ground, while supporting the EU Missions and positively impacting people's everyday lives. Launched by the European Commission in 2020, the programme aims to merge three key values—sustainability, inclusion, and aesthetics—to drive the Green Deal forward (https://new-european-bauhaus.europa.eu/index_en). It sustains a set of actions aligned with the EU Missions, from research to more practical and community-building initiatives, with the aim of integrating its key values into innovative solutions across various sectors that address climate, societal, and economic challenges.

The following Section presents the NEB as a program established to translate the European Green Deal into concrete changes in living environments that enhance the quality of the physical spaces, services, interactions, and experiences, positively impacting on people's daily lives. Social innovation projects and initiatives are framed as a fundamental component of the NEB, bringing a strong focus on collective actions on the ground that place social aspects at the center. The set of actions carried out by the NEB in its initial phase is described. The chapter concludes with a discussion of the progress of the NEB and the remaining challenges related to how the newly established NEB Facility can better leverage on social innovation for climate neutrality.

2   The New European Bauhaus: Tangible Transformations to Improve Everyday Lives

The New European Bauhaus programme was launched in September 2020 and formally established in 2021 to translate the European Green Deal into tangible improvements and concrete changes that enhance our everyday life in neighbourhoods, transforming living environments into sustainable, inclusive and beautiful places. The NEB addresses global challenges by focusing on local projects that offer a platform for experimentation while being a bridge between the worlds of science, technology, art and culture, and promotes a holistic approach to urban transformation, and promotes carbon–neutral lifestyles. These experiments focus on enhancing the quality and accessibility of spaces, restoring a sense of belonging, promoting life cycle thinking and circularity within industrial ecosystems, and reconnecting cities and people with nature through a multi-species, inclusive approach.

Taking inspiration from the Bauhaus experience, which lasted from 1919 to 1933 as a movement that had a profound influence on the future of architecture, design and the arts, the NEB aims to create a new movement to facilitate and steer the transformation of our living environments and societies according to three interconnected values: aesthetics (beautiful), sustainability (sustainable) and inclusion (together). *Beautiful* fosters creativity and social inclusion, emphasising collective experiences and values shaped by interactions between people and environments, including more-than-human perspectives and going beyond the functionality of places (von der Leyen 2020; Porta et al. 2016). *Sustainable* promotes minimizing environmental impact and regenerative design, which closes resource loops and fosters circular system through behavioural changes, education and participation (Gibbons 2020). *Together* emphasises the importance of collective action and cross-sector collaboration to design sustainable and inclusive environments, promoting social innovation and shared ownership of solutions by incorporating diverse perspectives of several groups (Rizzo et al. 2020; Giaccardi 2013). These values are operationalized through three guiding principles, which are shared with the EU Missions: participatory processes, multi-level engagement, and a transdisciplinary approach. *Participatory processes* and *transdisciplinary approaches* involve co-creation, empowering stakeholders to drive transformation through collective action and public sector reform (Voorberg et al. 2015; Puerari et al. 2017), where different actors contribute with their knowledge, capacities, and skills. *Multi-level engagement*, a key component of European policy, encourages collaboration across government and sector levels, fostering both horizontal and vertical cooperation between supranational, national, and sub-national levels, promoting partnerships and coordination (Hooghe and Marks 2001; Piattoni 2010). These principles guide the implementation of transformative NEB projects, aiming to foster inclusive and sustainable urban development (European Commission 2023). The following Section explores how the NEB programme has been implemented in its initial phase through different actions aimed at experimenting with the application of the aforementioned values and principles to improve the everyday lives of EU citizens while pursuing the key objectives of the Green Deal.

3   A Set of Initiatives Supporting the Implementation of the EU Green Deal and Missions

Throughout its implementation, the NEB programme has mobilized funds to support and promote innovative, transdisciplinary projects that embody its values and principles, contributing to the Green Deal through tangible demonstrators (https://eur opa.eu/rapid/press-release_IP-21-111_en.htm). The NEB was born directly from the political will of the Commission, when the future research and innovation programme was already set and thus had the need to find room within it and to connect research to more implementable initiatives with the aim of creating good practices and examples that could be followed. Therefore, the NEB has been advanced through the implementation of several projects under various EU funding calls that adopted a "marking and flagging approach". This approach entails identifying and labeling calls for proposals or projects across different EU programmes to ensure alignment with the core principles of the NEB, with Horizon Europe (HE) and the European Regional Development Fund (ERDF) serving as its main pillars (https://ec.europa.eu/ info/funding-tenders/opportunities/portal/screen/opportunities/ work-programmes). In 2021 and 2022, over €100 million in dedicated calls were allocated to support the NEB (https://new-european-bauhaus.europa.eu/index_en). Under Horizon Europe, the NEB promotes green, fair urban development and affordable housing, integrating with EU Cohesion Funds (2021–2027) and regional Partnership Agreements. A second generation of projects under the European Urban Initiative builds on the first wave efforts to enhance local authorities' transformation capacities. The NEB also fosters innovation in products, services and digital solutions, supporting small businesses through initiatives like the Worth Partnership Project II and the EIT Community NEB Booster, while cultural initiatives, such as the Culture Moves Europe and the European Year of Youth in 2022, embed NEB principles across Europe. Moreover, the NEB initiative has served as a connector, enhancing existing funding mechanisms and policy programmes (https://new-european-bauhaus.europa.eu/about/pro gress-report_en), inspiring several local, regional and national actors to create their own initiatives. Most of these initiatives are reported in the NEB dashboard (https:// web.jrc.ec.europa.eu/dashboard/NEB/), an evolving map of the NEB community. In this perspective, the reference to the original Bauhaus as a movement that was able to influence the future development of architecture and the arts, is intentional. The NEB's use of "lighthouse projects"[1] to spearhead its efforts aligns with this vision.

Despite a wide range of initiatives, the "marking and flagging approach" described above risked being inadequate to sustain the NEB's ambitious goals. In response, the European Commission in 2024 officially established the NEB Facility in 2024 (https://ec.europa.eu/newsroom/neb/items/837583/en). This cross-cutting, multi-year (2025–2027) initiative aims to integrate various funding instruments across the Commission. This programme provides a unique framework to revitalise European neighbourhoods by weaving Green Deal objectives with considerations

---

[1] CULTUURCAMPUS, NEB-STAR, NEBourhoods, DESIRE, EHHUR and Bauhaus of the Seas Sails.

of sustainability, inclusion and aesthetics. The NEB Facility comprises two main components—Research & Innovation (R&I) in Horizon Europe and Roll-out (https://new-european-bauhaus.europa.eu/new-european-bauhaus-facility_en). The R&I component focuses on researching, developing and testing innovative solutions to transform neighbourhoods through the NEB values. Supported by Horizon Europe, it is estimated to receive €120 million per year. The Roll-Out is dedicated to scaling, deploying and implementing the innovative solutions and business models developed under the R&I component. Its objectives include the delivery of NEB-inspired transformations by combining EU, public and private funding (https://new-european-bauhaus.europa.eu/get-inspired/selection-your-contributions/com mission-launches-consultations-2025-27-priorities-new-european-bauhaus-facility-2024-06-20_en). The Roll-out component builds on existing collaborations and develops new ones by harnessing common policy objectives between the NEB Facility and other relevant EU funds. This effort builds upon the work accomplished so far whereby the NEB has already mobilised eight EU programmes other than Horizon Europe.[2] Through its integrated approach, the NEB Facility aims to become more systemic and systematic than what the initial "marking and flagging" approach allowed for. It presents a well-structured general framework of components while allowing for adjustments and the inclusion of diverse initiatives. However, to ensure it achieves a meaningful impact on people's quality of life and a climate-neutral future, a few points of attention remain, which are addressed in the next Section.

4   The New European Bauhaus Supporting Social Innovation for Climate Neutrality: Progresses and Remaining Challenges

One of the key challenges for the NEB Facility is to be able to achieve its expected impacts while preserving an experimental approach and the capacity to build on the diversity of cultural backgrounds and institutional settings that can be found across Europe. In this perspective, the adoption of a portfolio-based approach emerges as a fundamental tool to develop diverse strategies to address transversal societal challenges in a variety of contextual conditions. Until now the ability to implement a portfolio-based approach has been limited due to the necessity of integrating NEB topics and calls into numerous existing programmes due to timing issues. This has resulted in a dilution of focus of the calls dedicated to the NEB, where resources and attention are spread too thinly across multiple areas, producing outcomes that risk being minimal. A portfolio-based approach would allow for a more cohesive and strategic allocation of resources, ensuring that each project receives the necessary attention and investment to maximise its potential. By focusing on a curated selection

---

[2] ERDF, LIFE, Single Market Programme, COSME, Digital Europe, Erasmus+, Creative Europe, European Solidarity Corps.

of initiatives, there can be a more meaningful engagement with key topics, resulting in stronger, more concentrated outcomes.

Another relevant challenge for the NEB Facility is to preserve its focus on the transformation of living environments without interpreting it only in terms of physical spaces and infrastructures. While physical redevelopment often garners more attention due to its visible, tangible results, it is the social dimension that can ensure the sustainability and success of these transformations. The way in which people perceive the quality of their living environments is indeed bound to many intangible factors and components (services, interactions, experiences, etc.) that need to be considered as much as the tangible components to achieve a satisfactory transformation towards climate neutrality. In this perspective, social aspects are fundamental: they are not only relevant components of the contextual conditions in which the transformation should occur, but also direct objectives of the transformation. Often overlooked, social, cultural and behavioural change and innovation are the most important factors that can make the green transition desirable, or at least acceptable. They can dramatically reduce the conflicts that the transition inevitably brings about and also directly contribute to achieving some of its objectives. By incorporating both physical and social elements, it is possible to foster more inclusive and holistic development processes that meet the needs of communities, not just in terms of infrastructure, but also in social cohesion, equity, and participation. In particular, social innovation can be a powerful means for mobilising common citizens, marginalised groups and different kinds of stakeholders, engaging them in initiatives that tackle concrete challenges and have at the same time the capacity to change the way in which they see problems and, ultimately, perceive the transition. Reflecting on the progress achieved, we must recognise that certain NEB actions focus on enhancing the capacities of regional and local authorities to support the development of social innovation projects that integrate the NEB values. By focusing on social innovation "on the ground", the NEB aligns with and exploits the concept of scaling deep, as described by Moore et al. (2015), where specific approaches are embedded within local contexts to create lasting change as a result of the operationalisation of the NEB values. Therefore, a more explicit reference to social innovation projects and initiatives could create environments that are not only physically improved but also socially enriched, supporting long-term positive change.

To conclude, it is essential to bring forward a reflection on the impact of the current approaches, methods and tools for monitoring, assessment and learning. With the transition from Horizon 2020 to Horizon Europe, the program logic recognised the difficulty of capturing the overall effects of small-scale initiatives and introduced the concept of "impact pathways". These pathways are designed during the proposal phase of each project and are subsequently implemented and refined to enhance the initial vision as the projects evolve. To achieve this, projects have been developing tailored monitoring and assessment frameworks, often incorporating elements of social innovation, such as in the case of NZC (Bresciani et al. 2024), and aligning with the support guidelines outlined in the NEB Compass (European Commission 2023). However, while NZC serves as the Mission Platform, it is only one of several efforts implemented under the Mission "Climate Neutral and Smart Cities" to support the Green Deal. A key question that remains is how to assess the cumulative impact of the European Commission's portfolio of actions under this Mission. Similarly, one of the most relevant challenges of the newly built NEB Facility lies in addressing the existing gap of a sufficiently wide range of evidence to convincingly demonstrate to EU policymakers across various levels of governance, as well as to the broader community of practitioners in the construction ecosystem, that the NEB's design values and principles are crucial for achieving the Green Deal's objectives, particularly in driving urban regeneration and transforming the built environment. For this, a specific action that aims at developing a monitoring and assessment framework tailored to the specific objectives of the NEB Facility is set to begin, with the twofold aim of informing the design of the pathways towards impact and the evaluation of the outcomes of the single projects, and the monitoring and assessment of the overall results and impacts of the whole programme.

Acknowledgements The work presented in this chapter was conducted as part of the projects:

"Desire—Designing the Irresistible Circular Society," which received funding from the European Union's Horizon Europe research and innovation programme under grant agreement No. 101079912. Website: https://www.irresistiblecircularsociety.eu.

"NetZeroCities" which received funding from the European Union's Horizon Europe research and innovation programme under grant agreement No. 101036519 (H2020) and No. 101121530—SGA-NZC (Horizon Europe) Website: https://netzerocities.eu/

The research was funded by the European Union. However, the views and opinions expressed are solely those of the authors and do not necessarily represent those of the European Union or CINEA. Neither the European Union nor the granting authority can be held responsible for them.

References

Bresciani S, Rizzo F, Mureddu F (2024) Assessment framework for people-centred solutions to carbon neutrality: a comprehensive list of case studies and social innovation indicators at urban level. Springer Nature, p 110

de Roo G (2018) Ordering principles in a dynamic world of change: on social complexity, transformation and the conditions for balancing intervention and spontaneity. Prog Plan 125:1–32

European Commission (2023) New European Bauhaus compass. Joint Research Centre

Franz H-W, Hochgerner J, Howaldt J (eds) (2012) Challenge social innovation: potentials for business, social entrepreneurship, welfare and civil society. Springer

Geels FW (2002) Technological transitions as evolutionary reconfiguration processes: a multi-level perspective and a case-study. Res Policy 31(8–9):1257–1274

Giaccardi E (2013) Metadesign for social innovation: enabling the multi-species future. J des Res 11(1):67–80

Gibbons LV (2020) Regenerative—the new sustainable? Sustainability 12(13):5483. https://doi.org/10.3390/su12135483

Healy J, Hughes J, Donnelly-Cox G, Shantz A (2024) A long and winding road: the hard graft of scaling social change in complex systems. J Bus Ventur Insights 2024(21):1–7

Hölscher K, Wittmayer JM, Loorbach D (2018) Transition versus transformation: what's the difference? Environ Innov Soc Trans 27:1–3

Hooghe L, Marks G (2001) Multi-level governance and European integration. Rowman & Littlefield

Mazzucato M (2018a) Mission-oriented innovation policies: challenges and opportunities. Ind Corp Chang 27(5):803–815

Mazzucato M (2018b) The entrepreneurial state: debunking public versus private sector myths. Penguin Books

Mazzucato M, Doyle S, Kimber N, Wainwright D, Wyld G (2024) Mission critical: Statecraft for the 21st century. UCL Institute for Innovation and Public Purpose, Policy Report 2024/04. UCL Institute for Innovation and Public Purpose

Moore M-L, Riddell D, Vocisano D (2015) Scaling out, scaling up, scaling deep: advancing systemic social innovation and the learning processes to support it. J Corp Citizsh 2015(58):67–84

Piattoni S (2010) The theory of multi-level governance: conceptual, empirical, and normative challenges. Oxford University Press

Porta S, Rofè Y, Vidoli M (2016) The city and the grid: building beauty at large scale

Puerari E, De Koning JIJC, Von Wirth T, Karré PM, Mulder IJ, Loorbach D (2017) Co-creation dynamics in urban living labs. Sustainability 10(6):1893

Rizzo F, Deserti A, Komatsu TT (2020) Implementing social innovation in real contexts. Int J Knowl Based Dev 11(1):45–67

UNDP (2022) System change: a guidebook for adopting portfolio approaches. Bangkok, Thailand

von der Leyen U (2020) State of the Union Address 2020. European Commission. https://ec.eur opa.eu/commission/presscorner/detail/en/SPEECH_20_1655

Voorberg WH, Bekkers VJJM, Tummers LG (2015) A systematic review of co-creation and co-production: embarking on the social innovation journey. Public Manag Rev 17(9):1333–1357